Lecture Notes
in Control and Information Sciences 211

Editor: M. Thoma

Springer-Verlag London Ltd.

Wodek Gawronski

Balanced Control of Flexible Structures

 Springer

Series Advisory Board

A. Bensoussan · M.J. Grimble · P. Kokotovic · H. Kwakernaak
J.L. Massey · Y.Z. Tsypkin

Author

Wodek Gawronski, Dr
Jet Propulsion Laboratory, California Institute of Technology, Pasadena,
California CA 91109, USA

ISBN 978-3-540-76017-7   ISBN 978-3-540-44435-0 (eBook)
DOI 10.1007/978-3-540-44435-0

British Library Cataloguing in Publication Data
A catalogue record for this book is available from the British Library

© Springer-Verlag London 1996
Originally published by Springer London Limited in 1996

Typesetting: Camera ready by author

69/3830-543210 Printed on acid-free paper

Preface

Control of flexible structures, a growing area of research, includes vibration suppression of bridges and tall buildings, active control of vehicle vibrations, suppression of rotor unbalanced dynamics, control of large flexible space structures, tracking control of antennas and radars, aircraft flutter suppression, and precision pointing of spaceborn reflectors.

This book provides analysis and design tools for the control of flexible structures. Flexible structures are linear systems of specific properties, most often known as modal properties. These distinctive properties make their analysis and controller design an unique task. Up to the present, the modal approach has been most successfully applied in the analysis of open-loop dynamics. However the modal approach is often used arbitrarily, since it lacks proper scaling properties (mode scale is a free factor, left to the designer discretion). For the closed-loop analysis and design, a balanced approach seems to be the most promising, since it combines modal properties with proper mode scaling, such that certain but crucial properties of a system (in this case controllability and observability) are reflected in the state equations. This book explains the balanced technique, derives the properties of flexible structures in the open- and closed-loop configuration, and presents a uniform approach to open- and closed-loop analysis and design.

The literature on the control of flexible structures is extensive. This short review will concentrate on books which deal with flexible structure control, or have a strong connection with the material presented in this book. Thus, our brief review of publications in the field of control of flexible structures is unavoidably biased.

Meirovitch has written what is already a classical book on control of flexible structures [88]. He develops models of flexible structures, and gives frequency and time domain methods of system analysis and synthesis. Porter [97] introduces modal control, which can be considered as a predecessor of balanced control. Skelton [105], although he treats control of general linear systems, undoubtedly introduces methods designed specifically for control of flexible structures. For example, the component cost approach to model or controller reduction is frequently used in the field. Junkins and Kim [66] give an up-to-date introduction to control of flexible structures; theirs is a graduate-level textbook on dynamics and control. Also, Junkins edited a monograph [65] that consists of up-to date contributions to the dynamics, identification, and control of flexible structures. Lin [78] gives a good and wide-ranging review of methods used in advanced system analysis and synthesis: H_2 and H_∞ controllers, robust design, nonlinear systems, fuzzy controllers, and control using neural networks. The monograph by Joshi [62] presents his own developments in the area of control flexible space structures, supported by numerous applications; it concentrates on robust dissipative controllers and LQG controllers. Inman's book [57] introduces the basic concepts of vibration control. Bryson's book [13] presents methods of spacecraft and aircraft control which includes, unavoidably, a flexible spacecraft control design methods. Maciejowski [84], and Doyle, Francis, and Tannenbaum [25] present the H_∞ control problems. Shahian and Hassul [100] give excellent introduction to the control design methods. Wortelboer [120] also reviews recent advances and presents new results in control design.

System identification is often a part of the controller design. "A good understanding of the connection between system identification and control is essential for success in the areas of vibration suppression ... control of flexible space structures, and control of robot manipulators" [64, p.xiv]. Efficient methods and software for flexible structure identification are given in the Juang's book [64]. Natke [92] and Ljung [80] present a wide range of identification methods useful in control engineering. Ewins [27] has written a text on practical modal testing of flexible structures. Gawronski and Natke [44] introduced balanced representation in the identification of flexible structures.

The reader is expected to have a basic background in control systems analysis. We do not provide this background, but to make the text more approachable, each presented result is extensively illustrated with examples and numerical data. The examples are twofold: simple ones that could be readily repeated by a reader, and complex ones (taken from the "real world") that demonstrate the effectiveness of the approach.

Chapter 1 presents the basic ideas and approaches which are taken further in the book. In Chapter 2 a flexible structure is defined and its properties specified. State-space representations and four forms of modal representations are introduced. Chapter 3 defines open-loop balanced systems and their properties. Specifically, properties of flexible structures in balanced coordinates derived in this chapter are further used for model order reduction and controller design. In Chapter 4 model reduction of an open-loop system is introduced. Optimality criteria of reduction show that in the case of flexible structures the reduction is near-optimal. Next, reduction of system with integrators, reduction in the finite time and/or frequency intervals, reduction based on transfer function measurements, and reduction of structures equipped with sensors and actuators are presented. The method is illustrated with the reduction of a COFS-1 structure (NASA Space Station test item), an Advanced Supersonic Transport, and a NASA Deep Space Network antenna. In Chapter 5 methods for sensor and actuator placement for balanced flexible structures are presented and illustrated using the actuator placement on the CEM large flexible structure at NASA Langley Research Center. In Chapter 6 the balanced identification procedure (ERA -Eigensystem Realization Algorithm) is introduced. ERA is described in detail in Ref.[64], and here is introduced for two reasons: it is a balanced approach to the system representation, and identification is tightly connected with the design of model based controllers. Chapter 7 introduces the balanced dissipative controllers. These controllers are simple and always stable, but their simplicity also creates difficulties: the performance depends on number of inputs and outputs (the more, the better), and for practical purposes, the collocation of inputs and outputs is required. In this chapter a formal design procedure of dissipative controllers for flexible structures is introduced. Chapter 8 defines a balanced LQG controller to be one for which the solutions of its Riccati equations are equal and diagonal. Properties of the LQG balanced systems are derived and later specified for flexible structures. These properties allow determination of the explicit relationship between the LQG weights and system performance. A method of reduction of the balanced LQG controller is introduced and illustrated with the LQG controller design for the NASA Deep Space Network antenna. In Chapter 9 a balanced H_∞ controller is introduced, and its properties that show its connection with the balanced LQG controller and balanced open-loop system are derived. This method of reduction of the H_∞ balanced controller gives stable reduced-order controllers with performance close to the full-order controller, as illustrated with controllers for truss structures and for the NASA Deep Space Network antenna. Finally, the Matlab® functions for open-loop balancing, as well as LQG and H_∞ balancing, are given in the Appendix B.

ACKNOWLEDGMENTS

My interest in flexible structures surfaced during my work with Jan Kruszewski at the Technical University of Gdansk, Poland. I appreciate his encouragement, inspiration, and enthusiasm, which led me into the fields of the finite element and structural analysis. Later, my research was directed into system identification when I joined Hans Guenther Natke at the University of Hanover, Germany. I acknowledge his initiative and hospitality. He introduced me to new topics and a new environment which so much changed my life. My knowledge of structural identification and control expanded when I joined Jer-Nan Juang at NASA Langley Research Center, Hampton, VA. He invited me into the aerospace community, where I enjoyed every day of work with him. Shyh J. Wang invited me to the Jet Propulsion Laboratory to work on space robot control problems. Soon, with Ben Parvin and Chris Yung I worked on the control issues of NASA's ground antennas, which gave me the opportunity to expand the knowledge of control of flexible structures, and check the results of analysis in implementation. Work with Ben and Chris is an interesting and inspiring experience.

I sincerely acknowledge the influence of my colleagues Fred Y. Hadaegh (Jet Propulsion Laboratory), Kyong B. Lim (NASA Langley Research Center), Jeff A. Mellstrom (Jet Propulsion Laboratory), and Trevor Williams (University of Cincinnati), who are co-authors of some methods included in this book, and whose cooperation and friendship I always enjoy.

I would like to thank Cornelius T. Leondes, editor of *Control and Dynamics Systems* series by Academic Press, who has shown a great deal of enthusiasm for my contributions to the series, and has encouraged me in my work. I am also obliged to my managers: Chris Yung, George Morris, and Joseph Statman for their support of the Deep Space Network antenna control study, part of which is included in this book.

I appreciate the effort of Jerzy T. Sawicki (Cleveland State University), who read the manuscript, and whose comments helped to improve it. The finite element model of Deep Space Network antenna was developed by Douglas Strain (Jet Propulsion Laboratory). I am obliged to Stephanie Nelson for voluntarily editing the manuscript.

Part of this research was performed at the Jet Propulsion Laboratory, California Institute of Technology, under contract with the National Aeronautics and Space Administration.

Finally, I thank my family - my wife Barbara, son Jacob, and daughter Kaja - for just being with me.

Wodek Gawronski

Pasadena, California
August 1995

Contents

Notation

General:

A^T: Transpose of matrix A

A^*:   Complex conjugate of the transpose of matrix A

A^{-1}:   Inverse of square nonsingular matrix A

$\|A\|$:   Norm of matrix A. Unless otherwise stated, it is the Euclidean norm, $\|A\| = \sqrt{\lambda_{max}(A^*A)}$

$diag(a_i)$:   Diagonal matrix with elements a_i along the diagonal

$exp(A)$:   Matrix exponential

$\mathcal{F}(.)$:   Fourier transformation operator

I_n: $n{\times}n$ unit matrix

rms   root-mean-square

t: Time, sec

T: Time interval

$tr(A)$:   Trace of square matrix A

$<x,y>$:   Scalar product of vectors x and y, $<x,y> = tr(xy^T)$

$\lambda(A)$:   Eigenvalue of square matrix A

ω: Circular frequency, rad/sec

Ω: Frequency interval; also matrix of modal frequencies

Linear systems:

(A,B,C,D):   System quadruple

H, G: Transfer functions

R: Transformation matrix

u: System input

x: System state

x_e: System estimated state

y: System output

Flexible systems:

D: Damping matrix

D_m: Modal damping matrix

K: Stiffness matrix; also controller gains

K_m: Modal stiffness matrix

M: Mass matrix

M_m: Modal mass matrix

q_i: i-th displacement of a flexible structure

q_{mi}: i-th modal displacement of a flexible structure

ϕ_i: i-th mode

Φ: Modal matrix

ω_i: i-th modal frequency

ζ_i: i-th modal damping

z: matrix of modal damping

Balanced open-loop systems:

$S(t)$ Time transformation matrix

$S(\omega)$ Frequency transformation matrix

W_c: Controllability grammian

W_o: Observability grammian

δ: Reduction index

ε: Optimality index

γ_i: i-th Hankel singular value

Γ: Diagonal matrix of the Hankel singular values

σ_i: i-th modal cost

Σ: Diagonal matrix of the modal costs

System identification:

$\mathcal{C}_p$: Controllability matrix of order p

$\mathcal{O}_p$ Observability matrix of order p

h_k Markov parameter (matrix)

H_1 Hankel matrix

H_2 Shifted Hankel matrix

p Number of samples

2

Δt	Sampling time
T	Time interval, $T=[0,\ p\Delta t]$
$W_c(p)$	Controllability grammian over time interval T
$W_o(p)$	Observability grammian over time interval T

LQG controllers:

CARE:	Controller algebraic Riccati equation
FARE:	Filter (estimator) algebraic Riccati equation
R:	LQG input weight matrix
q_i:	i-th weight of the LQG controller
Q:	LQG state weight matrix
S_c:	Solution of CARE
S_e:	Solution of FARE
v:	Process noise
V:	Process noise covariance matrix
w:	Measurement noise
W:	Measurement noise covariance matrix
β:	Pole shift coefficient
μ_i:	i-th LQG characteristic value
M:	Diagonal matrix of the LQG characteristic values
σ_i:	LQG controller reduction index
Σ:	Diagonal matrix of the LQG reduction indices

H_∞ controllers:

$\|G\|_\infty$:	H_∞ norm of the transfer function G
HCARE:	H_∞ controller algebraic Riccati equation
HFARE:	H_∞ filter (estimator) algebraic Riccati equation
$S_{\infty c}$:	Solution of HCARE
$S_{\infty e}$:	Solution of HFARE
$\mu_{\infty i}$:	i-th H_∞ characteristic value
M_∞:	Diagonal matrix of the H_∞ characteristic values
ρ:	HCARE and HFARE parameter
$\sigma_{\infty i}$:	H_∞ controller reduction index
Σ_∞:	Diagonal matrix of H_∞ reduction indices

Chapter 1

Introduction

In this book analysis and design tools for balanced control of flexible systems are developed. The term *flexible* refers to system properties, while the term *balanced* refers to a state of a system. It will be shown later that a combination of flexible systems and balanced states has special and useful properties that can be used to develop effective methods of control system analysis and design.

The term *flexible structure* is commonly used within the control engineering community to refer to a linear system with oscillatory properties that is characterized by a strong amplification of a harmonic signal for certain frequencies (called resonant frequencies), and whose transfer function poles are complex conjugate, typically with small real parts. Other attributes of flexible structures include weakly correlated states that correspond to a diagonal (modal) state matrix, or uncorrelated states that correspond to a diagonally dominant (balanced) state matrix.

Balanced system (and associated with it *balanced control*) has two meanings in this book. First, it is a system (or a control law) in a specific state-space representation such that its controllability and observability grammians are equal and diagonal. In this case the system is open-loop balanced. Second, it is a system (or a control law) in another state-space representation such that the solution of the linear quadratic Gaussian (LQG) or H_∞ Riccati equations are equal and diagonal. This representation is called LQG (or H_∞) balanced. For flexible structures, both balanced representations almost coincide, and, additionally, their components are almost independent. These two properties create a simple but powerful tool for open-loop model reduction, for the controller design, and for determination of the reduced-order LQG and H_∞ controllers.

This book tries to reflect the engineering approach to the controller design, which often relaxes analytical strictness in order to obtain implementable results. While strict mathematical analysis is necessary to obtain logically acceptable results, it does not warrant acceptable results in implementation. Consider, for example, a formal mathematical approach which guarantees controller properties, such as performance or stability. This assurance follows from the assumptions introduced to the analysis. However, the assumptions are more than often violated in implementation, causing the system to be unstable, or badly behaved. Thus, one has to keep in mind that analysis gives accurate results within the set of assumptions, and that implementation is the ultimate engineering test of usefulness.

The symbol "$\cong$" of approximate equality is used in this book. By using this approximate approach we were able to develop tools that would otherwise be not available. For example, this approach allows to design, in most cases, low-order controllers of required performance. The term "most cases" indicates that sometimes one can encounter a case which does not exactly have the expected properties. In this case a revised design is necessary. It is not a secret, however, that in the reality of our physical world nothing precisely satisfies assumptions associated with mathematical models of a physical process. Thus, engineers must expect to operate in this fuzzy environment, and make necessary adjustments to put theory to work. Consequently, approximation does not render an useless tool, nor unreliable results, but it rather makes the tools workable, and the results implementable. And this point is every so often demonstrated in this book. However, working in the approximate environment one has to check the results for their accuracy. In this book they have been checked both analytically and in implementation, for simple controllers and for complex ones (such as the controllers for the NASA Deep Space Network antennas). The new controllers, which were designed using the methods presented in this book, are now operational on several retrofitted NASA antennas, and their performance has been greatly improved. On implementation issues for these controllers see Ref.[46].

Chapter 2

Flexible Structures

The term *flexible structure* has different interpretations depending on source and application, and for the purpose of this book its meaning needs to be clarified. For flexible structures, as defined later, different state space representations are obtained. Among them modal representations are of great interest, and four modal state space representation of flexible structures are introduced. Two of them are especially useful in the further analysis.

2.1 Definition

A precise definition of the term "flexible structure" seldom surfaces in the control literature, where, most commonly, the definition is either omitted (as an obvious one), or narrowed to specific applications. The term is customarily used within the control engineering community to refer to a linear system with oscillatory properties that is characterized by a strong amplification of a harmonic signal for certain frequencies (called "resonant" frequencies), and whose transfer function poles are complex conjugate, typically with small real parts. This application-driven description of a flexible structure is commonly accepted, and no agreement about what properties make a system a flexible structure has been achieved. Indeed, it is often disputable whether a flexible structure is a linear system, or whether the real parts of its poles should be small (small damping property), or whether all its poles must be complex conjugate, etc.

For the purposes of this book, a flexible structure shall be defined as a finite-dimensional, controllable, and observable linear system with complex poles and with small damping. This definition narrows the class of linear systems under consideration, and based on this definition many interesting properties of structures and their controllers will be derived. However, our experience show that often, even if this definition of flexible structure is violated or extended, the derived properties still hold. For example, we will also consider unstable systems, heavy damping, some of the poles being real, etc. still obtaining reasonable results.

As a convention, we denote a dot as a first derivative with respect to time (i.e., $dx/dt=\dot{x}$), and double dot as a second derivative with respect to time (i.e., $d^2x/dt^2=\ddot{x}$). A linear system that fits the above definition of flexible structure is represented by the second-order matrix differential equation

$$M\ddot{q} +D\dot{q} +Kq=B_o u, \quad y=C_{oq}q+C_{ov}\dot{q} \qquad (2.1)$$

In this equation q is the $n_2 \times 1$ displacement vector, u is the $s \times 1$ input vector, y is the output vector, $r \times 1$, M is the mass matrix, $n_2 \times n_2$, D is the damping matrix, $n_2 \times n_2$, K is the stiffness matrix, $n_2 \times n_2$, the input matrix B_o is $n_2 \times s$, the output displacement matrix C_q is $r \times n_2$, and the output velocity matrix C_v is $r \times n_2$. The number n_2 is a number of degrees of freedom of the system (linearly independent coordinates describing the finite-dimensional structure), r is a number of outputs, and s is a number of inputs. The mass matrix is typically positive definite, and the stiffness and damping matrices are typically positive semidefinite. The above equation is obtained, for example, as a result of finite-element modeling, or as a result of a system identification.

2.2 State-Space Representations

For control analysis and design purposes, the flexible structure equation (2.1) is transformed into the state-space form

$$\dot{x}=Ax+Bu, \quad y=Cx \tag{2.2a}$$

The triple (A,B,C) is the state-space representation of a system, and x is the state vector. The state representation and the state vector are not unique, in the sense that different states give the same input-output relationship. For example, by introducing a new state variable, x_n, such that $x=Rx_n$ and R is nonsingular, a new state representation (A_n,B_n,C_n) is obtained

$$(A_n,B_n,C_n)=(R^{-1}AR, \ R^{-1}B,CR) \tag{2.2b}$$

This representation is equivalent to the original representation (A,B,C), since the input-output relationship has not been changed by the transformation. Although input-output relations remain invariant, it makes a difference what state representation is chosen for a controller design. Some representations have useful physical interpretations, others are more suitable for analysis and design. It is shown in this book that balanced and modal representations are specifically useful for the latter purpose.

In order to obtain a state representation for Eq.(2.1), it is rewritten (mass matrix is assumed nonsingular) as follows

$$\ddot{q} +M^{-1}D\dot{q} +M^{-1}Kq=M^{-1}B_o u, \quad y=C_q q+C_v\dot{q} \tag{2.3}$$

Define the state vector $x^T=[q^T \ \dot{q}^T]$, in which the first component is the system displacement, and the second component is the system velocity. In this case, after elementary manipulations, one obtains the following state-space representation

$$A = \begin{bmatrix} 0 & I \\ -M^{-1}K & -M^{-1}D \end{bmatrix}, \quad B = \begin{bmatrix} 0 \\ M^{-1}B_o \end{bmatrix}, \quad C = [C_q \ C_v] \qquad (2.4)$$

where A is $n{\times}n$, B is $n{\times}s$, and C is $r{\times}n$. The dimension of the state model n is twice the number of degrees of freedom of the system n_2, i.e., $n=2n_2$.

The order n of this representation is often unacceptably high; for example, the number of degrees of freedom of the finite-element model is typically very large. Therefore the latter state representation is seldom used in the engineering practice. An alternative approach is to represent Eq.(2.1) in a modal form using a modal matrix Φ $(n_2{\times}p)$, $p{\leq}n_2$, which consists of p eigenvectors (mode shapes) of a structure ϕ_i, $i=1,\dots,p$

$$\Phi = [\phi_1, \phi_2, \dots, \phi_p], \qquad p{\leq}n_2 \qquad (2.5)$$

The modes diagonalize mass and stiffness matrices M and K

$$M_m = \Phi^T M \Phi, \qquad K_m = \Phi^T K \Phi \qquad (2.6a)$$

obtaining diagonal matrices of modal mass M_m, stiffness K_m. These matrices are of dimension $p{\times}p$, and since the number of modes, p, is usually much smaller than the number of degrees of freedom, n_2, a substantial reduction in matrix dimension is achieved. If the damping matrix can be diagonalized as in Eq.(2.6a), it is called a matrix of proportional damping

$$D_m = \Phi^T D \Phi \qquad (2.6b)$$

The proportionality of damping is commonly assumed for analytical convenience, justified by the fact that the nature of damping is not exactly known, and thus its values are typically roughly approximated.

10

Introducing a new variable q_m (p×1), such that

$$q = \Phi q_m \tag{2.7}$$

and left-multiplying (2.1) by Φ^T, one obtains Eq.(2.3) in the form

$$\Phi^T M \Phi \ddot{q}_m + \Phi^T D \Phi \dot{q}_m + \Phi^T K \Phi q_m = \Phi^T B_o u, \qquad y = C_{oq} \Phi q_m + C_{ov} \Phi \dot{q}_m \tag{2.8a}$$

or equivalently

$$M_m \ddot{q}_m + D_m \dot{q}_m + K_m q_m = \Phi^T B_o u, \qquad y = C_{oq} \Phi q_m + C_{ov} \Phi \dot{q}_m \tag{2.8b}$$

hence

$$\ddot{q}_m + M_m^{-1} D_m \dot{q}_m + M_m^{-1} K_m q_m = M_m^{-1} \Phi^T B_o u, \qquad y = C_{oq} \Phi q_m + C_{ov} \Phi \dot{q}_m \tag{2.8c}$$

Denote $M_m^{-1} K_m = \Omega^2$, where $\Omega = diag(\omega_i)$ is a diagonal (p×p) matrix of natural frequencies ω_i (rad/sec), and $M_m^{-1} D_m = 2Z\Omega$, where $Z = diag(\zeta_i)$ is the modal damping matrix, ζ_i is the damping of the ith mode. With these notations the modal model is acquired from (2.8c)

$$\ddot{q}_m + 2Z\Omega \dot{q}_m + \Omega^2 q_m = M_m^{-1} \Phi^T B_o u, \qquad y = C_{oq} \Phi q_m + C_{ov} \Phi \dot{q}_m \tag{2.9}$$

Define the state variable $x^T = [x_1^T \ x_2^T] = [q_m^T \ \dot{q}_m^T]$, then (2.9) is presented as a set of first-order equations

$$\dot{x}_1 = x_2, \qquad \dot{x}_2 = -\Omega^2 x_1 - 2Z\Omega x_2 + M_m^{-1} \Phi B_o u, \tag{2.10a}$$

$$y = C_{oq} \Phi x_1 + C_{ov} \Phi x_2 \tag{2.10b}$$

or in the form (2.2a), where

$$A = \begin{bmatrix} 0 & I \\ -\Omega^2 & -2Z\Omega \end{bmatrix}, \quad B = \begin{bmatrix} 0 \\ M_m^{-1} \Phi^T B_o \end{bmatrix}, \quad C = [C_{oq} \Phi \ \ C_{ov} \Phi] \tag{2.11}$$

Thus Eq.(2.11) is the sought after state-space representation in modal coordinates. In this representation x_1 is the vector of the modal displacements, and x_2 is the vector of modal velocities. The dimension of this state-space representation is $2p$, while the state-space representation as in Eq.(2.4) is $2n_2$, and typically $2p \ll 2n_2$. While the mass and stiffness matrices are typically given in the "physical" coordinates from a finite element model, as matrices M and K in Eq.(2.1), the damping matrix is commonly evaluated in the modal coordinates, as a modal damping matrix Z. This is done because the damping estimation is more accurate in modal coordinates.

2.3 Modal Representations

Although the state-space representation presented above was obtained in the modal coordinates q_m, it is not a *modal state* representation. The modal state-space representation has a triple (A,B,C) in a special form, characterized by the block-diagonal matrix A

$$A=diag(A_i), \qquad B=\begin{bmatrix} B_1 \\ B_2 \\ \dots \\ B_{n2} \end{bmatrix}, \qquad C=[C_1,\ C_2,\ \dots,\ C_{n2}] \qquad (2.12)$$

$i=1,2,\dots,n_2$, where A_i, B_i, and C_i are 2×2, $2 \times s$, and $r \times 2$ blocks, respectively. The blocks A_i are in four different forms:

• modal form 1

$$A_i=\begin{bmatrix} 0 & \omega_i \\ -\omega_i & -2\zeta_i\omega_i \end{bmatrix} \qquad (2.13a)$$

12

- modal form 2

$$A_i = \begin{bmatrix} -\zeta_i\omega_i & \omega_i \\ -\omega_i & -\zeta_i\omega_i \end{bmatrix}$$ (2.13b)

- modal form 3

$$A_i = \begin{bmatrix} 0 & 1 \\ -\omega_i^2 & -2\zeta_i\omega_i \end{bmatrix}$$ (2.13c)

- modal form 4

$$A_i = \begin{bmatrix} -\zeta\omega_i + j\omega_i\delta_i & 0 \\ 0 & -\zeta_i\omega_i - j\omega_i\delta_i \end{bmatrix},$$ (2.13d)

where $j = \sqrt{-1}$, and $\delta_i = \sqrt{1 - \zeta_i^2}$. The ith state component, x_{ik}, corresponding to the ith block of the kth modal form are as follows

$$x_{i1} = \begin{bmatrix} q_{mi} \\ \dot{q}_{mi}/\omega_i \end{bmatrix}, \quad x_{i2} = \begin{bmatrix} q_{mi} \\ q_{moi} \end{bmatrix}, \quad x_{i3} = \begin{bmatrix} q_{mi} \\ \dot{q}_{mi} \end{bmatrix}, \quad x_{i4} = \begin{bmatrix} q_{mi} - jq_{moi} \\ q_{mi} + jq_{moi} \end{bmatrix}$$ (2.14)

where q_{mi}, $\dot{q}_{mi}$ are ith modal displacement and velocity, as defined in Eq.(2.7), and $q_{moi} = \zeta_i q_{mi} + \dot{q}_{mi}/\omega_i$.

The first three are real modal representations, while the last one is a complex modal representation. In this book the first two forms are used extensively, in contrast to the last form, where a complex matrix A_i obviously creates unnecessary numerical difficulties.

The modal form 3 is obtained from the representation (A,B,C) as in Eq.(2.11) by a simple rearranging of columns of A and C and rows of A and B. In this way the state vector $x^T = [q_m^T \ \dot{q}_m^T]$, consisting of modal

13

displacements followed by modal rates, is transformed to the new state $x_n^T = [q_{m1} \ \dot{q}_{m1} \ q_{m2} \ \dot{q}_{m2} \ \cdots \ q_{mn2} \ \dot{q}_{mn2}]$, where modal displacement for each component stays next to its rate. This is done by using the transformation matrix R in the form

$$R = \begin{bmatrix} 0 & e_1 \\ e_1 & 0 \\ 0 & e_2 \\ e_2 & 0 \\ \cdots & \cdots \\ \cdots & \cdots \\ 0 & e_{n2} \\ e_{n2} & 0 \end{bmatrix}, \quad (2.15)$$

where e_i is a n_2 row vector with all elements equal to zero except the ith which is equal to one.

Denote A_k as the state matrix A in the modal form k, where $k=1,2,3,$ or 4. The transformation matrix, R_{kl}, transforms the modal state variable x_k into x_l, $x_l = R_{kl}x_k$, $k,l=1,2,3,$ or 4

$$R_{kl} = diag(R_{kli}) \quad (2.16)$$

and assuming small damping, i.e. $\zeta_i \ll 1$, $i=1,\ldots,n_2$, one obtains

$$R_{12i} = \begin{bmatrix} 1 & 0 \\ \zeta_i & 1 \end{bmatrix}, \quad R_{13i} = \begin{bmatrix} 1 & 0 \\ 0 & \omega_i \end{bmatrix}, \quad R_{14i} = \begin{bmatrix} 1-j\zeta_i & -j \\ 1+j\zeta_i & j \end{bmatrix} \quad (2.17a)$$

$$R_{23i} = \begin{bmatrix} 1 & 0 \\ -\zeta_i\omega_i & \omega_i \end{bmatrix}, \quad R_{24i} = \begin{bmatrix} 1 & -j \\ 1 & j \end{bmatrix}, \quad R_{34i} = \begin{bmatrix} 1-j\zeta_i & -j/\omega_i \\ 1+j\zeta_i & j/\omega_i \end{bmatrix} \quad (2.17b)$$

The remaining transformations can be easily derived from the above ones by using $R_{kpi} = R_{pki}^{-1}$, or by noting that $R_{kpi} = R_{lpi}R_{kli}$, $l,k,p=1,2,3,4$. The Matlab subroutine *cdf2rdf.m* transforms the modal form 4 into the modal form 2. The subroutine *realdiag.m* given in the Appendix B1 transforms any state-space representation into the modal representation 2.

2.4 Examples

In this book different types of flexible structures are investigated, but three types are used most frequently: a simple structure, a truss, and the Deep Space Network (DSN) antenna. They represent different levels of complexity.

2.4.1 Simple Structure

A three-mass system - a simple flexible structure - is used mainly for the illustration purposes. Its simplicity allows for easy analysis, and straightforward interpretation. Also in this case, solution properties and numerical data can be displayed in a compact form. The system is shown in Fig.2.1, with masses m_1, m_2, and m_3, stiffness k_1, k_2, k_3, and k_4, and a damping matrix which is a linear combination of stiffness and mass matrices, $D=\alpha K+\beta M$, where $\alpha>0$, $\beta>0$ are constants.

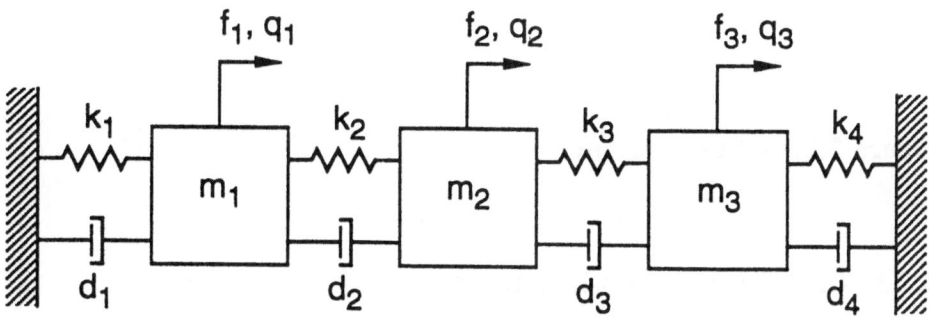

Figure 2.1: A simple flexible structure

2.4.2 Truss

A truss structure is another example of a flexible structure which is closer to real structures, yet can be easily simulated by a reader, if necessary. The structure is presented in Fig.2.2. For this structure $d_1=70$ in., $d_2=100$ in., each truss has a cross-section area of 2 in.^{2}, an elastic modulus of 10^6 lb/in.^{2}, and mass density of 2 lb sec^2/in.^{2}. Vertical control forces are applied at nodes $n7$ and $n8$, and the output rates are measured in the vertical direction at nodes $n3$ and $n4$. The system has 26 states (13 components), two inputs, and two outputs. Its stiffness and mass matrices are given in Appendix A.

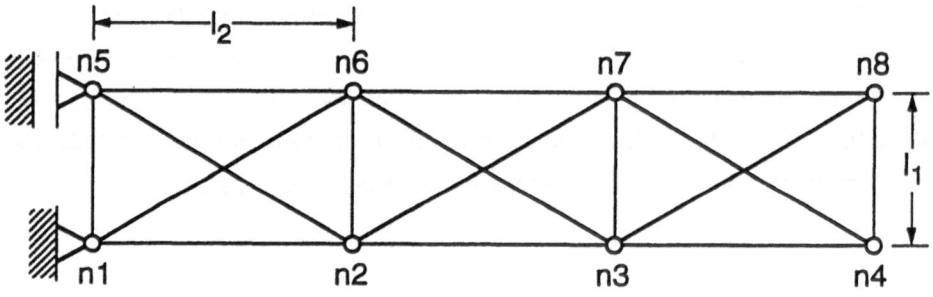

Figure 2.2: A truss structure

2.4.3 Deep Space Network Antenna

The NASA Deep Space Network antenna structure illustrates a real-world complex flexible structure. The Deep Space Network antennas, operated by the Jet Propulsion Laboratory, consist of several antenna types and

Figure 2.3: DSS-13 - the Deep Space Network antenna

are located at Goldstone (California), Canberra (Australia), and Madrid (Spain). The Deep Space Network serves as a communication tool for space exploration. A new generation Deep Space Network antenna with the 34-meter dish is shown in Fig.2.3. This antenna is an articulated large flexible structure, which can rotate around azimuth and elevation axes. The rotation is controlled by azimuth and elevation servos, as in Fig.2.4. The configuration of the antenna structure and its azimuth and elevation drives is the open-loop (or rate-loop) model of the antenna. The open loop plant has two-input and two-outputs, and the position

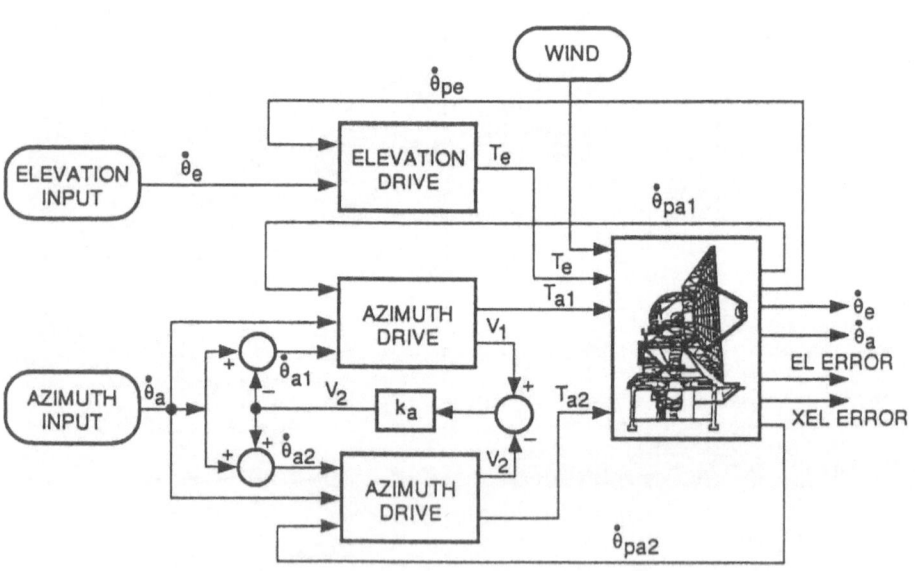

Figure 2.4: The rate loop model of the Deep Space Network antenna

18

loop is closed between the encoder outputs, and the rate inputs. The drives consist of gearboxes, electric motors, amplifiers, and tachometers. For more details about the antenna see Ref.[43]. It is a large structure (number of degrees of freedom $n_2 = 2000$ or more) which has some nonlinear properties (dry friction, backlash, stick-slip friction, limits on rates and acceleration). This large and complex structure presents many difficulties in the controller design, and also violates the narrow definition of a flexible structure provided earlier. Yet, its flexible properties are dominating that the controller design methods derived for the narrowly defined structure can be applied in the antenna case as well.

Chapter 3

Balanced Open-Loop Systems

Controller design relies heavily on the plant controllability and observability properties. Heuristically, controllability is a property of a plant input to excite the total system dynamics (i.e., a property that all states are excited by the input actuators). Observability, on the other hand, is a property of an output and includes total system dynamics (i.e., the property that all states are sensed by the output sensors). Individually, neither controllability nor observability provide sufficient information for a controller design. For example, if parts of the states are weakly controllable (weakly influenced by the plant input), it is impossible to decide on controller configuration, since the same states could be strongly observable by the output sensors. Similarly, the weakly observable states, which are weakly reflected in the plant output, could be strongly excited by the input actuators. However, if parts of the states are weakly controllable and observable at the same time (weakly influenced by the plant input and weakly recorded in the plant output), they can be ignored in the controller design. The joint controllability and observability properties of the system are characterized by the Hankel singular values, and this simple idea was developed by Moore [89].

Instead of using the controllability and observability properties, Skelton [101], [102], [103], [105], [106] characterizes each state variable with its cost. The cost of a state variable reflects its participation in the system output. If the participation is strong, the cost is high, and *vice versa*. Two drawbacks should be noted briefly. First, cost evaluates a state at the system output, while the input seems to be irrelevant. Second, if two states are highly coupled, their individual cost does not reflect their actual participation in the output. The first drawback is removed by implementing either white noise or impulse input. These inputs excite system dynamics equally at all frequencies. The second drawback not always can removed. Fortunately, in balanced or modal representation of flexible structures, components are weakly coupled, hence their cost can be easily evaluated.

3.1 Observability and Controllability Grammians

The controllability and observability grammians are the convenient form of characterization of a system controllability and observability properties. They are defined in the following. Consider a linear, observable and controllable system with state-space representation (A,B,C). The system is assumed stable unless otherwise stated. It can be, of course, a flexible structure, but not necessarily. Let λ_i be the ith eigenvalue of A and impose the condition $\lambda_i+\lambda_k\neq0$ for every $i,k=1,\dots,n$. The controllability and observability grammians are defined as follows [67]

$$W_c(t)=\int_0^t exp(A\tau)BB^T exp(A^T\tau)d\tau, \quad W_o(t)=\int_0^t exp(A^T\tau)C^T C exp(A\tau)d\tau \quad (3.1)$$

Alternatively, they can be determined from the following differential equations

$$\dot{W}_c=AW_c+W_cA^T+BB^T, \quad \dot{W}_o=A^TW_o+W_oA+C^TC \quad (3.2)$$

The stationary solutions of Eq.(3.2) are obtained in the limiting case of $t \to \infty$. In this case, for a stable system one obtains $\dot{W}_c=\dot{W}_o=0$, and

21

the grammians are determined from the following Lyapunov equations

$$AW_c + W_cA^T + BB^T = 0, \quad A^TW_o + W_oA + C^TC = 0 \tag{3.3}$$

For stable A the solutions of Eq.(3.3) are positive definite

The grammians depend on the system coordinates, and for a linear transformation of a state, $\bar{x} = Rx$, they are transformed as follows

$$\overline{W}_c = R^{-1}W_cR^{-T}, \quad \overline{W}_o = R^TW_oR \tag{3.4}$$

However, the eigenvalues of the grammian product are invariant under linear transformation, since

$$\lambda_i(\overline{W}_c\overline{W}_o) = \lambda_i(R^{-1}W_cR^{-T}R^TW_oR) = \lambda_i(R^{-1}W_cW_oR) = \lambda_i(W_cW_o) \tag{3.5}$$

This property is often used later.

3.2 Time-Limited Grammians

The stationary grammians are defined from the Lyapunov equations (3.3). The grammians defined over finite time interval $[0, t)$ can be also determined from the Lyapunov equations, rather then from the differential equations (3.2). In order to show it assume that a system is excited, and the response is measured within the time interval $T = [t_1, t_2]$, $\infty > t_2 > t_1 \geq 0$. According to Eq.(3.1) the grammians over the time interval T are defined as follows

$$W_c(T) = \int_{t_1}^{t_2} exp(A\tau)BB^T exp(A^T\tau)d\tau, \quad W_o(T) = \int_{t_1}^{t_2} exp(A^T\tau)C^TC exp(A\tau)d\tau \tag{3.6}$$

Note that for stable A these grammians are positive definite: $W_c(T) > 0$,

$W_o(T) > 0$ if $t_2 > t_1$. These grammians can be obtained from the stationary grammians W_c, W_o as follows (see Refs.[37], [38])

$$W_c(T) = W_c(t_1) - W_c(t_2), \qquad W_o(T) = W_o(t_1) - W_o(t_2) \qquad (3.7)$$

where

$$W_c(t) = S(t) W_c S^T(t), \qquad W_o(t) = S^T(t) W_o S(t), \qquad (3.8)$$

and

$$S(t) = -exp(At) \qquad (3.9)$$

where W_c, W_o are unlimited time grammians, i.e., the solutions of Eq.(3.3).

In order to prove it, note that from Ref.[67] we have $W_c([o,t]) = W_c - W_c(t)$, but $W_c(T) = W_c([0,t_1]) - W_c([0,t_2]) = W_c(t_1) - W_c(t_2)$. The observability grammian $W_o(T)$ is obtained similarly.

3.3 Band-Limited Grammians

The grammians can be defined and interpreted in the frequency domain. According to the Parseval theorem, the integrals in Eq.(3.1) over the time interval $[0, \infty)$ can be replaced with the frequency domain integrals

$$W_c = \frac{1}{2\pi} \int_{-\infty}^{\infty} H(\nu) BB^T H^*(\nu) d\nu, \qquad W_o = \frac{1}{2\pi} \int_{-\infty}^{\infty} H^*(\nu) C^T CH(\nu) d\nu \qquad (3.10)$$

where

$$H(\nu) = (j\nu I - A)^{-1} \qquad (3.11)$$

23

is a Fourier transformation of *exp(At)*.

The grammians in Eq.(3.10) are defined over the entire frequency domain, i.e., for $\omega \in (-\infty, \infty)$. In the following, the band-limited grammians, or grammians in the finite frequency domain $\Omega=[\omega_1, \omega_2]$, $\infty > \omega_2 > \omega_1 \geq 0$ are introduced. It will be shown that they are also determined from the Lyapunov equations. In order to obtain grammians in the frequency band Ω, assume that the impulse responses pass through a filter with frequency window $[\omega_1, \omega_2]$, so that from Eq.(3.10) one obtains

$$W_c(\Omega)=W_c(\omega_2)-W_c(\omega_1), \quad W_o(\Omega)=W_o(\omega_2)-W_o(\omega_1) \tag{3.12}$$

where

$$W_c(\omega)= \frac{1}{2\pi} \int_{-\omega}^{\omega} H(v)BB^TH^*(v)dv, \quad W_o(\omega)= \frac{1}{2\pi} \int_{-\omega}^{\omega} H^*(v)C^TCH(v)dv \tag{3.13}$$

Note again that $W_c(\omega)$ and $W_o(\omega)$ are positive definite. By using the Lyapunov equations (3.3) and noting that $-AW_c-W_cA^T=H^{-1}(v)W_c+W_cH^{-T}(v)$, Eq.(3.13) becomes

$$W_c(\omega)= \frac{1}{2\pi} \int_{-\omega}^{\omega} (H(v)W_c+W_cH^*(v))dv \tag{3.14}$$

which produces

$$W_c(\omega)=W_cS^*(\omega)+S(\omega)W_c \tag{3.15}$$

where

24

$$S(\omega) = \frac{1}{2\pi} \int_{-\omega}^{\omega} H(\nu)d\nu = \frac{j}{2\pi} \ln((A+j\omega I)(A-j\omega I)^{-1}) \qquad (3.16)$$

and W_c is the controllability grammian obtained from Eq.(3.3). Thus, in Eq.(3.15) the controllability grammian in a finite frequency interval is obtained from the stationary controllability grammian by implementing the frequency weighting $S(\omega)$.

Similarly, the observability grammian in the finite frequency interval is determined

$$W_o(\omega) = S^*(\omega)W_o + W_o S(\omega) \qquad (3.17)$$

where W_o is the stationary observability grammian obtained from the Lyapunov equations (3.3).

Finally, for the frequency interval $\Omega = [\omega_1, \ \omega_2]$, $\omega_2 > \omega_1$, the grammians are obtained by combining Eqs.(3.12), (3.16), and (3.17)

$$W_c(\Omega) = W_c S^*(\Omega) + S(\Omega)W_c \quad W_o(\Omega) = S^*(\Omega)W_o + W_o S(\Omega) \qquad (3.18)$$

where $S(\Omega) = S(\omega_2) - S(\omega_1)$. Note that $H(\nu_1)$, $H(\nu_2)$ commute for any ν_1, and ν_2, i.e., $H(\nu_1)H(\nu_2) = H(\nu_2)H(\nu_1)$. From this property, and definitions (3.16), (3.11) one obtains

$$S(\omega)W_c = \frac{1}{4\pi^2} \int_{-\omega}^{\omega}\int_{-\infty}^{\infty} H(\nu)H(\mu)BB^{\mathrm{T}}H^*(\mu)d\mu d\nu$$

$$= \frac{1}{2\pi} \int_{-\infty}^{\infty} H(\mu)S(\omega)BB^{\mathrm{T}}H^*(\mu)d\mu \qquad (3.19)$$

and from Eq.(3.15)

$$W_c(\omega) = W_c S^*(\omega) + S(\omega) W_c = \frac{1}{2\pi} \int_{-\infty}^{\infty} H(\mu) Q_c(\omega) H^*(\mu) d\mu \tag{3.20}$$

where

$$Q_c(\omega) = S(\omega) B B^{\text{T}} + B B^{\text{T}} S^*(\omega) \tag{3.21}$$

For a stable matrix A, the integral (3.20) can be obtained by solving the following Lyapunov equation

$$A W_c(\omega) + W_c(\omega) A^{\text{T}} + Q_c(\omega) = 0 \tag{3.22}$$

and consequently from Eq.(3.18), the grammian $W_c(\Omega)$ is computed from the Lyapunov equation

$$A W_c(\Omega) + W_c(\Omega) A^{\text{T}} + Q_c(\Omega) = 0 \tag{3.23}$$

where

$$Q_c(\Omega) = Q_c(\omega_2) - Q_c(\omega_1) \tag{3.24}$$

Similarly, the observability grammian $W_o(\Omega)$ is obtained by solving the following Lyapunov equation

$$A^{\text{T}} W_o(\Omega) + W_o(\Omega) A + Q_o(\Omega) = 0 \tag{3.25}$$

where

$$Q_o(\Omega) = Q_o(\omega_2) - Q_o(\omega_1), \quad Q_o(\omega) = S^*(\omega) C^{\text{T}} C + C^{\text{T}} C S(\omega) \tag{3.26}$$

3.4 Time- and Band-Limited Grammians

The time- and band-limited grammians are the grammians in the combined finite time and frequency intervals. They are determined by limiting them either first in time and then frequency domain, or vice versa. The results are identical in both cases, since the frequency and time domain operators commute, as shown below.

Consider the controllability grammian in the finite time interval, defined in Eq.(3.6) and determined from Eq.(3.7). According to the Parseval theorem, the grammian in the infinite time interval $[0 \; \infty)$ is represented in the infinite frequency domain $(-\infty, \; \infty)$ from Eq.(3.13). Assuming that $H_c(\omega) = H(\omega)B$ in Eq.(3.13) is measured inside the finite frequency interval $[-\omega, \; \omega]$, $W_c(\omega)$ can be determined from Eq.(3.15). Introducing Eq.(3.15) to Eq.(3.8) yields

$$W_c(t,\omega) = S(t)W_c(\omega)S^*(t) \tag{3.27}$$

where $S(t)$ is given by Eq.(3.8) and $W_c(\omega)$ by Eq.(3.15).

Consider now the controllability grammian in the frequency domain and apply the Parseval theorem in the time domain to obtain

$$W_c(t,\omega) = W_c(t)S^*(\omega) + S(\omega)W_c(t) \tag{3.28}$$

where $W_c(t)$ is given in Eq.(3.8), and $S(\omega)$ in Eq.(3.16). The question arises whether $W_c(t,\omega)$ in Eqs.(3.27) and (3.28) are identical. In order to show that it is true, note that $S(t)$ and $S(\omega)$ commute

$$S(t)S(\omega) = S(\omega)S(t) \tag{3.29}$$

for any ω and t. Indeed, since

$$S(\omega) = \int_{-\omega}^{\omega} (j\nu I - A)^{-1} d\nu = \int_{-\omega}^{\omega} \mathcal{F}(e^{A\tau}) d\nu \qquad (3.30)$$

where $\mathcal{F}(.)$ denote the Fourier transformation operator, one obtains

$$S(t)S(\omega) = \int_{-\omega}^{\omega} e^{At} \mathcal{F}(e^{A\tau}) d\nu = \int_{-\omega}^{\omega} \mathcal{F}(e^{A\tau}) e^{At} d\nu = S(\omega)S(t) \qquad (3.31)$$

Next, introduction of Eq.(3.15) to Eq.(3.27), with the aid of Eq.(3.29), yields

$$W_c(t,\omega) = S(t)W_c S^T(t)S^*(\omega) + S(\omega)S(t)W_c S^T(t) \qquad (3.32)$$

Recalling Eq.(3.8), the equality of (3.27) and (3.28) is thus proved.

As a consequence of the commuting property, the controllability grammian over the finite time interval $T=[t_1,\ t_2]$, $t_2 > t_1$, and the frequency interval $\Omega=[\omega_1,\ \omega_2]$, $\omega_2 > \omega_1$ is determined from

$$W_c(T,\Omega) = W_c(T,\omega_2) - W_c(T,\omega_1) = W_c(t_1,\Omega) - W_c(t_2,\Omega) \qquad (3.33)$$

where

$$W_c(T,\omega) = W_c(t_1,\omega) - W_c(t_2,\omega), \qquad W_c(t,\Omega) = W_c(t,\omega_2) - W_c(t,\omega_1) \qquad (3.34)$$

and

$$W_c(t,\omega) = S(t)W_c(\omega)S^*(t) = W_c(t)S^*(\omega) + S(\omega)W_c(t) \qquad (3.35)$$

Similarly, the observability grammian over the time interval T and frequency interval Ω is obtained from

$$W_o(T,\Omega) = W_o(T,\omega_2) - W_o(T,\omega_1) = W_o(t_1,\Omega) - W_o(t_2,\Omega) \qquad (3.36)$$

where

$$W_o(T,\omega)=W_o(t_1,\omega)-W_o(t_2,\omega), \quad W_o(t,\Omega)=W_o(t,\omega_2)-W_o(t,\omega_1) \qquad (3.37)$$

and

$$W_o(t,\omega)=S^*(t)W_o(\omega)S(t)=W_o(t)S(\omega)+S^*(\omega)W_o(t) \qquad (3.38)$$

3.5 Systems with Integrators

Some plants (for example plants of tracking systems) include integrators. A linear system is a system with integrators if its $n-m$ poles are stable, the remaining m poles are at zero, and it is observable and controllable (see [23]). It is assumed also that A is nondefective (c.f. [54]), i.e., the geometric multiplicity of poles at zero is m. Grammians for systems with integrators do not exist; but, systems with integrators are controllable and observable, nevertheless. These properties can be still measured quantitatively, although in an alternative way [42], by noting that for systems with integrators, inverses of the grammians exist.

The inverses of the grammians, called antigrammians, are defined as follows. For a controllable and observable triple (A,B,C), the matrices V_c, V_o satisfying the Riccati equations

$$V_cA+A^TV_c+V_cBB^TV_c=0, \quad V_oA^T+AV_o+V_oC^TCV_o=0 \qquad (3.39)$$

are the controllability and observability antigrammians. For stable controllable and observable systems $V_c=W_c^{-1}$, $V_o=W_o^{-1}$, where W_c, W_o are controllability and observability grammians and satisfy the Lyapunov equations (3.3).

For a system with integrators one obtains $V_c=diag(0_m \; V_{co})$, $V_o=diag(0_m \; V_{oo})$, where 0_m is a $m \times m$ zero matrix. In order to prove this, consider (A,B,C) in the form

$$A=\begin{bmatrix} 0_m & 0 \\ 0 & A_o \end{bmatrix}, \quad B=\begin{bmatrix} B_r \\ B_o \end{bmatrix}, \quad C=[C_r \ C_o] \tag{3.40}$$

Matrix A in the form (3.40) always exists due to m poles at zero, and B, C exist due to the nondefectiveness of A. From Eq.(3.39) it follows that

$$V_c B_r=0, \quad C_r V_{or}=0, \quad V_{cro}=0, \quad V_{oro}=0 \tag{3.41}$$

where V_c, V_o are divided conformably to A:

$$V_c=\begin{bmatrix} V_{cr} & V_{cro} \\ V_{cro}^T & V_{co} \end{bmatrix}, \quad V_o=\begin{bmatrix} V_{or} & V_{oro} \\ V_{oro}^T & V_{oo} \end{bmatrix} \tag{3.42}$$

For a controllable and observable system the matrices B_r and C_r are of full rank, thus it follows from Eq.(3.14) that $V_{cr}=0$, $V_{or}=0$, and that $V_c=diag(0_m \ V_{co})$, $V_o=diag(0_m \ V_{oo})$.

3.6 Balanced Representation

The system triple is open-loop balanced, if its controllability and observability grammians are equal and diagonal, as defined by Moore in Ref.[89]

$$W_c=W_o=\Gamma^2, \quad \Gamma=diag(\gamma_1,...,\gamma_n), \quad \gamma_i \geq 0, \quad i=1,...,n \tag{3.43}$$

where γ_i is the ith Hankel singular value of the system. The transformation R to the balanced representation, such that $(A_b,B_b,C_b)=(R^{-1}AR, \ R^{-1}B, \ CR)$, is determined as follows, see Ref.[38]

$$R=PU\Gamma^{-1}, \quad R^{-1}=\Gamma^{-1}V^TQ \tag{3.44}$$

The matrices Γ, V, U are obtained from the singular value decomposition of the matrix H

$$H = V\Gamma^2 U^{\mathsf{T}} \tag{3.45}$$

where H is obtained as a product of the matrices P and Q

$$H = QP \tag{3.46}$$

and P, Q are obtained from the decomposition of the controllability and observability grammians respectively

$$W_c = PP^{\mathsf{T}}, \qquad W_o = Q^{\mathsf{T}}Q \tag{3.47}$$

This decomposition can be achieved by the Cholesky, or singular value decomposition.

The Matlab function *bal_op.m* which transforms a representation (A,B,C) to the open-loop balanced representation (A_b, B_b, C_b) is given in the Appendix B2.

3.7 Balanced Systems with Integrators

First, note that for a controllable, observable, and balanced system, its antigrammians are equal and diagonal

$$V_c = V_o = \Pi^2 \tag{3.48}$$

where $\Pi = diag(\pi_i)$, $i=1,2,\ldots,n$, and satisfies the following equations

$$\Pi^2 A_b + A_b^{\mathsf{T}} \Pi^2 + \Pi^2 B_b B_b^{\mathsf{T}} \Pi^2 = 0, \qquad \Pi^2 A_b^{\mathsf{T}} + A_b \Pi^2 + \Pi^2 C_b^{\mathsf{T}} C_b \Pi^2 = 0 \tag{3.49}$$

If a system with integrators is balanced, one obtains $\Pi = diag(0_m \ \Pi_o)$, where 0_m is a $m \times m$ zero matrix. In this case A_b is block-diagonal,

31

$A_b = diag(0_m \ A_{bo})$.

A balanced representation of a system with integrators (A_b, B_b, C_b) is obtained by the transformation R

$$A_b = R^{-1}AR, \quad B_b = R^{-1}B, \quad C_b = CR \tag{3.50}$$

where the transformation R is a combination of the two subsequent transformations, R_1 and R_2

$$R = R_1 R_2 \tag{3.51}$$

The transformation R_1 turns A into the block-diagonal form $A_1 = diag(0_m, A_o)$ (e.g. into modal form)

$$A_1 = R_1^{-1}AR_1 \tag{3.52}$$

and the transformation R_2 is in the form $R_2 = diag(I_m, R_o)$ where I_m is an identity matrix of order m, and R_o balances A_o

$$A_{bo} = R_o^{-1}A_o R_o \tag{3.53}$$

which can be easily checked by introducing Eq.(3.41) to Eq.(3.49).

3.8 Balanced Flexible Structures

The controllability and observability properties of flexible structures are analyzed in this chapter. Most of them can be expressed in the closed form, and some of them were already presented in Refs.[10], [38], [47], [51], [61], [102], [115], and [118].

Two important properties of a flexible structure are extensively used in this book. First, that the states of a flexible structure in modal coordinates are almost orthogonal, and second, that the state matrix A in the balanced coordinates is diagonally dominant. Both terms, "almost orthogonal" and "diagonally dominant," are alternative expressions for an approximate equality. The latter term is used in the following sense. Two variables, x and y, are approximately equal ($x \cong y$) if $x = y + \varepsilon$, and $\|\varepsilon\| \ll \|y\|$. For example, if M is a diagonal matrix, then S is diagonally dominant, ($S \cong M$) if the diagonal terms s_{ii} and μ_{ii} satisfy the condition $s_{ii} + \varepsilon = \mu_{ii}$, and the off-diagonal terms of S satisfy the condition $|s_{ij}| < \varepsilon$, where ε is small, and $i,j = 1,...,n$.

3.8.1 Principal Properties

Assuming small damping, such that $\zeta \ll 1$ ($\zeta = max(\zeta_i)$, $i = 1,...,n_2$), the balanced and modal representations of flexible structures are closely related, as it is expressed in the following properties:

Property 1. *Diagonally dominant grammians in modal coordinates.* In modal coordinates controllability and observability grammians are diagonally dominant, i.e.,

$$W_c \cong diag(w_{ci}I_2), \qquad w_{ci} > 0, \quad i = 1,...,n_2 \qquad (3.54a)$$

$$W_o \cong diag(w_{oi}I_2), \qquad w_{oi} > 0, \quad i = 1,...,n_2 \qquad (3.54b)$$

and therefore their product is also diagonally dominant. The eigenvalues of the product are the Hankel singular values at fourth power, see Eq.(3.5), thus denote $W = W_c W_o$, and the approximate Hankel singular value as the diagonal term of W, $\gamma^4_{iapprox} = w_{ii}$, to obtain

$$\gamma^4_i \cong \gamma^4_{iapprox} = w_{ii} \qquad (3.55)$$

where Γ is as in Eq.(3.43).

The controllability grammian is, at the same time, a covariance matrix of the states x excited by the white noise input, i.e., $W_c = E(xx^T)$. Thus, its diagonal dominance corresponds to the almost independent, or almost orthogonal states.

Property 2. *The "conservation law" for the approximate Hankel singular values.* The sum of the approximate Hankel singular values in fourth power (as defined in Eq.(3.55)) is equal to the sum of the exact Hankel singular values in fourth power

$$\sum_{i=1}^{n} \gamma_i^4 = \sum_{i=1}^{n} \gamma_{i\,approx}^4 \qquad (3.56)$$

This property shows that if some of the approximate Hankel singular values are larger than the exact ones, then the other approximate Hankel singular values are smaller than the exact ones, such that Eq.(3.56) holds. This property can serve as a tool for checking the correctness of the approximation. For example, if all the approximate Hankel singular values are larger than the exact ones, the approximation cannot be considered correct.

Property 3. *Balancing by rescaling the modal coordinates.* By rescaling the modal representation (A_m, B_m, C_m) one obtains almost balanced representation. Let (A_b, B_b, C_b) be the balanced representation, then

$$(A_b, B_b, C_b) \cong (A_m, R_{mb}^{-1} B_m, C_m R_{mb}) \qquad (3.57)$$

where R_{mb} is a diagonal matrix

$$R_{mb} = diag(R_{mbi}), \qquad R_{mbi} = r_i\, I_2, \qquad i = 1, \dots, n_2 \qquad (3.58a)$$

and

$$r_i = \left(\frac{w_{ci}}{w_{oi}}\right)^{1/4} \qquad\qquad (3.58b)$$

Note that the transformation matrix R_{mb} leaves matrix A_m almost unchanged, and scales matrices B_m and C_m.

Property 4. *Balanced modes and natural modes.* A flexible structure in modal coordinates is described by its *natural modes*, ϕ_i, $i=1,\dots,n_2$, and in the balanced representation it is described by the *balanced modes*, ϕ_{bi}, $i=1,\dots,n_2$. The latter ones are obtained by re-scaling natural modes

$$\phi_{bi} \cong r_i \; \phi_i, \qquad i=1,\dots,n_2, \qquad\qquad (3.59)$$

where the scaling factor r_i is given by Eq.(3.58b).

Property 5. *Diagonally dominant system matrix in the balanced coordinates.* For a balanced flexible structure, its matrix A is diagonally dominant

$$A \cong diag(A_i), \qquad i=1,2,\dots,n_2 \qquad\qquad (3.60)$$

and A_i has either modal form 1, see Eq.(2.13a), or modal form 2, see Eq.(2.13b).

Property 6. *The relationship between A, B, and C matrices in the balanced coordinates.* For balanced representation in modal form 1, one obtains

$$BB^T \cong C^T C \cong -\Gamma^2 (A+A^T) = diag(0,2\alpha_1,0,2\alpha_2,\dots,0_n,2\alpha_n), \qquad (3.61a)$$

or, for the *i*th block

35

$$B_i B_i^T \cong C_i^T C_i \cong -\gamma_i^2 (A_i + A_i^T) = \begin{bmatrix} 0 & 0 \\ 0 & 2\alpha_i \end{bmatrix} \qquad (3.61b)$$

and for the balanced representation in modal form 2 one arrives at

$$BB^T \cong C^T C \cong -\Gamma^2(A + A^T) = diag(\alpha_1, \alpha_1, \alpha_2, \alpha_2, \ldots, \alpha_n, \alpha_n), \qquad (3.61c)$$

which, for the ith block, it translates to

$$B_i B_i^T \cong C_i^T C_i \cong -\gamma_i^2 (A_i + A_i^T) = \alpha_i I_2 \qquad (3.61d)$$

where $\alpha_i = 2\gamma_i^2 \zeta_i \omega_i$, B_i is the two-row block of B, and C_i is the two-column block of C.

In order to prove the first property consider a flexible structure in the modal representation 1, as in Eqs.(2.12) and (2.13a). The diagonal dominance of the controllability and observability grammians can be proved by introduction of this modal representation to the Lyapunov equations (3.3). By inspection, for $\zeta \to 0$ one obtains the finite values of the off-diagonal terms w_{cij}, and w_{oij}, for $i \neq j$, i.e., $\lim_{\zeta \to 0} w_{cij} < \infty$, and $\lim_{\zeta \to 0} w_{oij} < \infty$, while the diagonal terms tend to infinity, $\lim_{\zeta \to 0} w_{cii} = \infty$, and $\lim_{\zeta \to 0} w_{oii} = \infty$. Moreover, the difference between the diagonal terms in each block is finite, $\lim_{\zeta \to 0} |w_{cii} - w_{ci+1,i+1}| < \infty$, thus for small ζ the grammians in modal coordinates are diagonally dominant, having 2×2 blocks on the diagonal with almost identical diagonal entries of each block. Eq.(3.55) is a direct consequence of the diagonal dominance of W_c and W_o, and the fact that the eigenvalues of the grammian product are invariant, see Eq.(3.5).

The second property is proved as follows

$$\sum_{i=1}^{n} \gamma_i^4 = \sum_{i=1}^{n} \lambda_i(W_cW_o) = tr(W_cW_o) = tr(W) = \sum_{i=1}^{n} \gamma_{i\,approx}^4 \qquad (3.62)$$

The first equality follows from the definition of the Hankel singular values, the second one is the following property of the square matrix X, $\sum_{i=1}^{n} \lambda_i(X) = tr(X)$, and the third one from Eq. (3.55).

The third property is proved by inspection. Introducing Eqs. (3.58) and (3.54) to Eq. (3.4) one obtains

$$\overline{W}_c \cong \overline{W}_o \cong \Gamma^2 = diag(\gamma_i^2 I_2), \qquad \gamma_i = (w_{ci}w_{oi})^{1/4} \qquad (3.63)$$

hence, the balanced representation (A_b, B_b, C_b) is obtained from the modal representation (A_m, B_m, C_m) by means of Eq. (3.57).

A similar property can be derived for the modal representation 2. The closeness of these two modal representations follows from the fact that the transformation R_{12} from modal representation 1 to 2 is itself diagonally dominant, as in Eq. (2.17a).

The fourth property is proven as follows. For the modal matrix Φ, as in Eq. (2.5) one obtains from Eq. (2.7) that

$$\begin{bmatrix} q \\ \dot{q} \end{bmatrix} = \begin{bmatrix} \Phi & 0 \\ 0 & \Phi \end{bmatrix} \begin{bmatrix} q_m \\ \dot{q}_m \end{bmatrix} = \begin{bmatrix} \Phi & 0 \\ 0 & \Phi \end{bmatrix} \begin{bmatrix} x_{m1} \\ x_{m2} \end{bmatrix} \qquad (3.64a)$$

or, equivalently

$$q = \sum_{i=1}^{p} \phi_i\, x_{m1i}, \qquad \dot{q} = \sum_{i=1}^{p} \phi_i\, x_{m2i} \qquad (3.64b)$$

37

But, from Eq.(3.58) it follows that $x_{m1i}=r_i x_{b1i}$, and $x_{m2i}=r_i x_{b2i}$, thus Eq.(3.64b) is now

$$q=\sum_{i=1}^{p} r_i \phi_i\, x_{b1i} = \sum_{i=1}^{p} \phi_{bi}\, x_{b1i}, \qquad \dot{q}=\sum_{i=1}^{p} r_i \phi_i\, x_{b2i} = \sum_{i=1}^{p} \phi_{bi}\, x_{b2i} \qquad (3.65)$$

where ϕ_{bi} is a balanced mode as in Eq.(3.59).

The fifth property is proven by noting that the balanced matrix A_b is similar to the modal matrix A_m, i.e., $A_b=R_{mb}^{-1}A_m R_{mb}$, where R_{mb} is as in Eq.(3.58). Since A_m is block-diagonal, and R_{mb} is block-diagonally dominant, thus A_b is also block-diagonally dominant.

The last property follows from the diagonal dominance of the matrix A in the balanced coordinates. Namely, by introducing Eqs.(3.60) and (3.43) to the Lyapunov equations (3.3), one obtains (3.61).

The above properties show that the balanced representation of flexible structures is almost independent of the actuator and sensor location (in other words, matrix A in the balanced representation is almost invariant, as in Eq.(3.60), regardless of the sensor and actuator location), and that the orientations of the balanced and modal coordinates are almost identical. Although the balanced and modal coordinates almost coincide, the important difference between them lies in their scaling. It is well known that the modal coordinates are not unique, since they depend on the scaling of the natural modes, which is arbitrary. For example, the scaling of the natural modes is chosen to obtain a unit modal mass matrix, or unit length modes. The scaling of the balanced coordinates is unique, it is such that the condition (3.61) holds.

The above properties, which connect A, B, and C, and the grammians, are used later to derive additional properties of balanced flexible structures and controllers.

3.8.2 Closed-Form Grammians in Modal Coordinates

In order to further investigate the connections between the flexible structures in the balanced and modal coordinates, a closed-form grammians are derived. Consider an n_2-mode model for a flexible structure with low damping, i.e., such that $\zeta_i < 1$, and separated natural frequencies, ω_i, $\omega_i \neq \omega_j$, for $i \neq j$, and $i,j=1,...,n$. Assume also that if the natural frequencies occur in clusters of repeated frequencies, the modes related to these frequencies are mutually independent. Note that ω_i and ζ_i are positive numbers. Consider also the modal equations as in Eq.(2.9)

$$\ddot{q}_m + diag(2\zeta_i\omega_i)\dot{q}_m + diag(\omega_i^2)q_m = Bu, \quad y = C_r\dot{q}_m + C_d q_m \qquad (3.66)$$

where q_m is the vector of modal coordinates, $q = \Phi q_m$. Defining the state vector $x^T = [\omega_1 q_{m1}, \quad \dot{q}_{m1}, \quad ..., \quad \omega_{n2} q_{mn2}, \quad \dot{q}_{mn2}]$, the state-space representation in this case is of the first form, as in (2.13a)

$$A_i = \begin{bmatrix} 0 & \omega_i \\ -\omega_i & -2\zeta_i\omega_i \end{bmatrix}, \quad B_i = \begin{bmatrix} 0 \\ b_i \end{bmatrix}, \quad C_i = [c_{di}/\omega_i \quad c_{ri}] \qquad (3.67)$$

where b_i is the ith row of B and c_{ri} c_{di} are the jth columns of C_r and C_d, respectively.

In order to determine the closed-form grammians define the scalars

$$\beta_{ij} = tr(B_i B_j^T) = b_i b_j^T, \qquad (3.68a)$$

$$\theta_{ij} = tr(C_i^T C_j) = c_{ri} c_{rj} + \frac{c_{di} c_{dj}}{\omega_i \omega_j} \qquad (3.68b)$$

The closed form controllability grammian is derived in Refs.[106], [115], and [47], which provide

$$W_{cij} = \frac{\beta_{ij}}{d_{ij}} \times \begin{bmatrix} 2\omega_i\omega_j(\zeta_j\omega_i + \zeta_i\omega_j) & \omega_j(\omega_j^2 - \omega_i^2) \\ -\omega_i(\omega_j^2 - \omega_i^2) & 2\omega_j\omega_i(\zeta_i\omega_i - \zeta_j\omega_j) \end{bmatrix} \qquad (3.69a)$$

where $d_{ij} = 4\omega_i\omega_j(\zeta_i\omega_i + \zeta_j\omega_j)(\zeta_j\omega_i + \zeta_i\omega_j) + (\omega_j^2 - \omega_i^2)^2$. For $i=j$ one obtains

$$W_{cii} = \frac{\beta_{ii}}{4\zeta_i\omega_i} \times I_2 \qquad (3.69b)$$

This expression for W_{cij} is simplified considerably for separated frequencies and for ζ_i, $\zeta_j \rightarrow 0$. In this case,

$$W_{cij} \rightarrow \begin{bmatrix} 0 & \omega_j \\ -\omega_i & 0 \end{bmatrix} \times \frac{\beta_{ij}}{\omega_j - \omega_i} \qquad (3.70)$$

Thus, W_c for a flexible structure with light damping and separated frequencies has the ith diagonal block proportional to $1/\zeta_i$ and off-diagonal blocks independent of ζ_i, therefore it tends to a diagonal form for $\zeta_i \rightarrow 0$, as it was shown earlier.

The observability grammian for a system with rate measurements only can be obtained in a similar fashion, or by noting that $A^T = PAP$ where P is a signature matrix, $P = diag(1,-1,\ldots,1,-1)$. Pre- and post-multiplying the Lyapunov equations (3.3) by P gives

$$A(PW_oP) + (PW_oP)A^T + C^TC = 0 \qquad (3.71)$$

where the identity $CP = C$ was used. Thus, W_{oij} is similar to W_{cij}, where the latter is given by Eq.(3.69). The difference is in the signs of the off-diagonal entries, and β_{ij} is replaced by $\theta_{rij} = c_{ri}^T c_{rj}$.

If displacement measurements are also allowed, the closed-form expressions for W_o are too complex to be useful. The only exception is the expression for the ith diagonal block of W_o, where

$$W_{oii} \cong I_2 \times \frac{\omega_i^2 \theta_{rii} + \theta_{dii}}{4 \zeta_i \omega_i^3} = I_2 \times \frac{\theta_{ii}}{4 \zeta_i \omega_i}, \tag{3.72a}$$

where

$$\theta_{ii} = \frac{\theta_{dii}}{\omega_i^2} + \theta_{rii}, \qquad \theta_{dii} = c_{di}^T c_{di} \tag{3.72b}$$

Although no analytical expressions for W_o are viable, it is possible to evaluate this matrix which using the modal form 1 of (A,B,C). Let

$$W_{oij} = \begin{bmatrix} p & q \\ r & s \end{bmatrix} \tag{3.73}$$

be the (i,j)th block of W_o. Then its Lyapunov equation can be rewritten as the system of linear equations

$$\begin{bmatrix} 2(\zeta_i \omega_i + \zeta_j \omega_j) & -\omega_j & -\omega_i & 0 \\ \omega_j & 2\zeta_i \omega_i & 0 & -\omega_i \\ \omega_i & 0 & 2\zeta_j \omega_j & -\omega_j \\ 0 & \omega_i & \omega_j & 0 \end{bmatrix} \begin{bmatrix} p \\ q \\ r \\ s \end{bmatrix} = \begin{bmatrix} c_{ri}^T c_{rj} \\ c_{ri}^T c_{dj}/\omega_j \\ c_{di}^T c_{rj}/\omega_i \\ c_{di}^T c_{dj}/\omega_i \omega_j \end{bmatrix} \tag{3.74}$$

For this equation the determinant of the left-hand matrix in Eq.(3.74) is d_{ij} as in Eq.(3.69), so this quantity plays a similar role in both the controllability and observability grammians.

For the grammians as in the above equations (3.69b) and (3.72) one can find the Hankel singular values γ_i as the diagonal entry of $W_{cii} W_{oii}$

$$\gamma_i^2 \cong \frac{\sqrt{\beta_{ii} \theta_{ii}}}{4 \zeta_i \omega_i}, \tag{3.75}$$

41

which confirms the result given by Gregory [51]. This expression is used to obtain modal costs for a flexible structure, which are the diagonal elements of $-2\Lambda W_c W_o \cong -2\Lambda diag(\gamma_i^4)$. For the modal representation in form 2, and from Eq.(3.75), one obtains the cost

$$\sigma_i = \gamma_i^2 \sqrt{2\zeta_i\omega_i} = \sqrt{\frac{\beta_{ii}\theta_{ii}}{8\zeta_i\omega_i}} \tag{3.76}$$

for the ith mode (modal cost of Skelton [106]).

Example 3.1   The exact Hankel singular values, obtained from the algorithm in Section 3.6, and the approximate Hankel singular values from Eq.(3.75) are compared for the simple structure as in Fig.2.1. For this system with masses $m_1 = m_2 = m_3 = 1$, stiffness $k_1 = k_2 = k_4 = 6$, $k_3 = 20$, and a damping matrix as was given in the modal form, $z = diag(0.005, 0.005, 0.005)$, there are two inputs, forces at the mass 2 and 3, and two outputs, displacement of masses 1 and 3.

The exact and the approximate Hankel singular values, compared in Table 3.1, show good coincidence.

Table 3.1   Hankel singular values of the simple system

State no	γ_{exact}	γ_{approx}
1	3.0009	2.9934
2	2.9860	2.9934
3	1.3153	1.3121
4	1.3087	1.3121
5	0.8682	0.8661
6	0.8639	0.8661

Example 3.2   The exact and the approximate Hankel singular values are compared for the truss as in Fig.2.2. Its outputs are rates measured at vertical direction at nodes $n4$ and $n8$, respectively. The inputs are: (1) member force in the bar connecting node $n2$ and the basis, (2)

42

member force in the bar connecting node *n3* and *n2*, (3) memeber force in the bar connecting node *n4* and *n3*, (4) member force in the bar connecting node *n6* and the basis, (5) force in the bar connecting node *n7* and *n6*, and (6) force in the bar connecting node *n8* and *n7*.

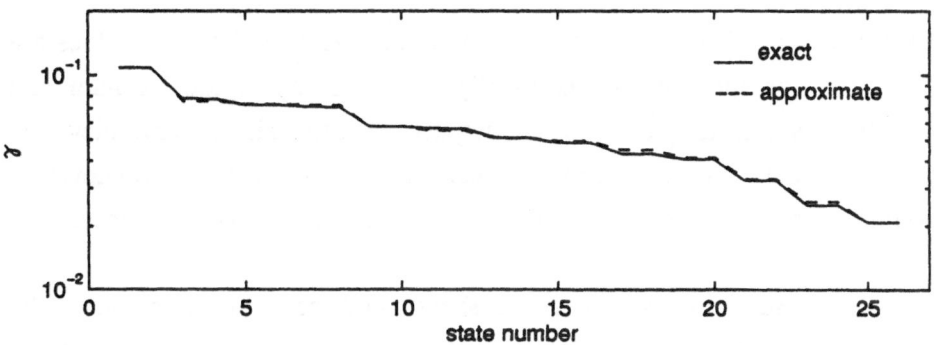

Figure 3.1: Exact and approximate Hankel singular values for the truss.

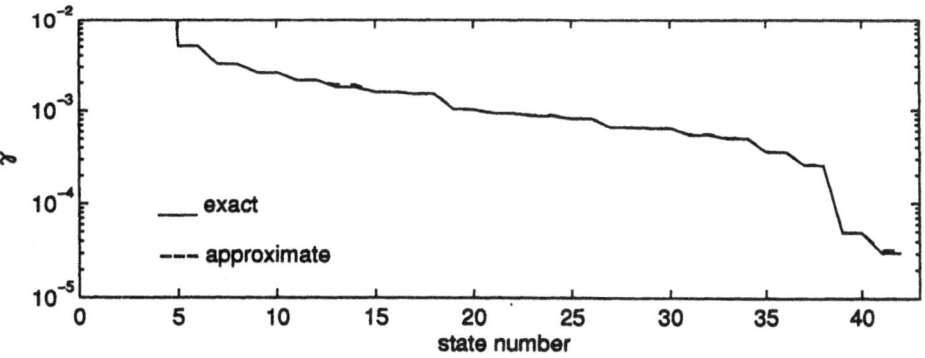

Figure 3.2: Exact and approximate Hankel singular values for the antenna.

The plots in Fig.3.1 of the exact and approximate Hankel singular values show that the approximation error is insignificant, and that the "conservation law," as in Property 2, applies: the Hankel singular value overestimation is followed by the underestimation, such that the sum of exact and approximate Hankel singular values at fourth power is the same, as in Eq.(3.56).

Example 3.3 The exact and the approximate Hankel singular values are compared for the antenna as in Fig.2.3. The inputs are azimuth and elevation pinion torques, the 16 outputs include azimuth and elevation position and rates, pinion rates, elevation and cross-elevation pointing errors, and subreflector position in three-dimensional space.

The results in Fig.3.2 again show good coincidence between the exact and the approximate Hankel singular values. The first four Hankel singular values correspond to the antenna rigid body motion in azimuth and elevation and are not shown in this figure, since their values tend to infinity. The treatment of the rigid body case was described in Section 3.5.

Chapter 4

Model Reduction

Model reduction is a part of the dynamic analysis of flexible structures. Typically, a model with a large number of degrees of freedom, such as the one developed for the static analysis of structures, causes numerical difficulties in the dynamic analysis, to say nothing of the high computational cost. Additionally, if one takes into account that the complexity of a controller depends on the plant order then it is not difficult to see that a full-order controller for a high-order plant is hardly implementable. Thus the reduction of system order solves the problem assuming that the reduced model acquires essential properties of the full-order model.

Model reduction has a quite a long history, and many reduction techniques have been developed. Some reduction methods, as in Refs.[55], [116], and [117], give optimal results, but they are complex and computationally expensive. Another methods, comparatively simple, give results close to the optimal one. The latter include balanced and modal truncation, see Refs.[89], [51], [102], [105], [106], [38], and [47], which are discussed later. Model reduction of flexible structures requires specific approach, and it was studied by Gregory [51], Jonckheere [61], Skelton, [105] and Gawronski [38], Gawronski and Williams [47], and Cottin, Prells, and Natke [18]. This chapter is based partially on Ref.[38].

4.1 Reduction Technique and Reduction Indices

A reduced-order model of a linear system is obtained here by truncating appropriate states in the balanced or modal coordinates. In order to do this let the state vector x of the system in the balanced representation be partitioned as follows, $x^T = [x_r^T, \; x_t^T]$, and the state triple (A,B,C) be partitioned accordingly,

$$A = \begin{bmatrix} A_r & A_{rt} \\ A_{tr} & A_t \end{bmatrix}, \quad B = \begin{bmatrix} B_r \\ B_t \end{bmatrix}, \quad C = [C_r \; C_t] \tag{4.1}$$

where x_r is a $k \times 1$ vector and x_t is an $(n-k) \times 1$ vector. The balanced grammian is divided similarly,

$$\Gamma^2 = \begin{bmatrix} \Gamma_r^2 & 0 \\ 0 & \Gamma_t^2 \end{bmatrix} \tag{4.2}$$

The reduced model is obtained by deleting the last $n-k$ rows of A,B, and the last $n-k$ columns of A,C, i.e.,

$$A_r = LAL^T, \quad B_r = LB, \quad C_r = CL^T, \quad L = [I_k \; 0] \tag{4.3}$$

In order to make the right choice of truncated components the reduction index, Δ is introduced

$$\Delta = \|y - y_r\| \tag{4.4}$$

When the reduction index is normalized with respect to the system output introduction of the normalized reduction index δ is obtained

$$\delta = \frac{\|y - y_r\|}{\|y\|} = \frac{\Delta}{\|y\|} \tag{4.5}$$

Typically, although not necessarily, $\delta \ll 1$ for near-optimal reduction.

The index Δ is determined from the formula

$$\Delta^2 = \|y\|^2 + \|y_r\|^2 - 2tr(BB_r^T W_{cr}^2) \qquad (4.6)$$

where $W_{cr} = E(x_r x^T)$ is the covariance between the states of the original and the reduced model, and $E(.)$ is the expectation operator. On determination of W_{cr} see Ref.[36]. It is seen from Eq.(4.6) that the minimum of Δ cannot be determined, unless the full combination of truncated states is exhausted (the difficulty lies in the determination of the covariance W_{cr}). This inconvenience is avoided if the indices Δ and δ are replaced with the indices Δ_e and δ_e, which are the estimated reduction errors defined as follows

$$\Delta_e^2 \cong \|y\|^2 - \|y_r\|^2, \qquad \delta_e^2 \cong 1 - \frac{\|y_r\|^2}{\|y\|^2} \qquad (4.7)$$

In the above equation $\|y\|^2 = tr(C^T C \Gamma^2)$, and $\|y_r\|^2 = tr(C_r^T C_r \Gamma_r^2)$.

Define a scalar product of vectors x and y as $<x,y> = tr(xy^T)$. Considering the case of $\delta \ll 1$, one obtains

$$<y_r, y> \cong <y_r, y_r> \qquad (4.8)$$

and from the definition (4.4) and property (4.8)

$$\Delta^2 = \|y - y_r\|^2 = \|y\|^2 + \|y_r\|^2 - 2<y, y_r> \cong \|y\|^2 + \|y_r\|^2 - 2<y_r, y_r> = \|y\|^2 - \|y_r\|^2 \quad (4.9)$$

Thus for the near-optimal reduction, the indices Δ and Δ_e, as well as δ and δ_e, are nearly identical

$$\Delta_e \cong \Delta, \qquad \delta \cong \delta_e \qquad (4.10)$$

The distance of the reduced model from the hypothetical optimal one is evaluated from the necessary optimality condition, see Ref.[83], which states that y_r must be orthogonal to the error $y - y_r$. This condition can be written as $<y_r, y-y_r> = 0$. Thus the optimality index ε

$$\varepsilon = \frac{|<y_r,\ y-y_r>|}{\|y_r\|\,\|y-y_r\|} \qquad (4.11)$$

is a measure of the distance of the reduced model from the optimal one. Surely, for the optimal reduced model one obtains $\varepsilon = 0$, and for the near-optimal model one obtains $\varepsilon \ll 1$.

Although the optimality index ε relates the reduced model to the optimal reduced model, in application it can be replaced with the reduction index δ as an indicator of optimality. Indeed, note that for $\delta \ll 1$ Eq.(4.8) holds, thus $<y_r, y-y_r> = <y_r, y> - <y_r, y_r> \cong 0$, and $\varepsilon \cong 0$. In summary, $\delta \ll 1$ implies $\varepsilon \ll 1$, but the opposite is not true. However, the latter situation of $\varepsilon \ll 1$ and δ close to 1 is not interesting from the application point of view; even if the reduction is near-optimal, the reduction should be avoided due to the large truncation error.

4.2 Near-Optimal Reduction in Balanced Coordinates

In this section, conditions for near-optimal reduction in the balanced coordinates are presented. Denote c_i the ith column of C, b_i the i-th row of B, a_{ii} the i-th diagonal element of A, then the following holds

$$c_i^T c_i = b_i b_i^T = -2\gamma_i^2 a_{ii} \qquad (4.12a)$$

Denote the above value as α_i^2

$$\alpha_i^2 = c_i^T c_i = b_i b_i^T = -2\gamma_i^2 a_{ii} \qquad (4.12b)$$

and define the component cost σ_i of the ith state, similarly to Refs.[102], [105], which represents the contribution of the state variable x_i in the output

$$\sigma_i^2 = \alpha_i^2 \gamma_i^2, \tag{4.13}$$

This definition follows from the output norm in the balanced coordinates

$$\|y\|^2 = trE(yy^T) = trE(C^T Cxx^T) = tr(C^T C\Gamma^2) = \sum_{i=1}^{n} \alpha_i^2 \gamma_i^2 = \sum_{i=1}^{n} \sigma_i^2 \tag{4.14}$$

thus σ_i indicates the participation of x_i in the output y.

Furthermore, define matrix Σ

$$\Sigma^2 = diag(\sigma_i^2), \quad i=1,\dots,n \tag{4.15}$$

and denote $\Sigma = diag(\Sigma_r, \Sigma_t)$. Note that in open-loop balanced representation $\|y\|^2 = tr(\Sigma^2)$, $\|y_r\|^2 = tr(\Sigma_r^2)$, thus

$$\Delta_e^2 = \|y\|^2 - \|y_r\|^2 = tr(\Sigma^2) - tr(\Sigma_r^2) = tr(\Sigma_t^2) \tag{4.16}$$

as well as

$$\delta_e^2 = \frac{tr(\Sigma_t^2)}{tr(\Sigma^2)} \tag{4.17}$$

Since $tr(\Sigma^2)$ is fixed, it is easy to see that the smallest δ_e is obtained for the smallest Σ_t. Thus, in order to obtain a near-minimal reduction error, rearrange the coordinates such that the matrix Σ has its diagonal entries in decreasing order, i.e., $\sigma_i \geq \sigma_{i+1} > 0$, $i=1,\dots,n\text{-}1$. The approximate value of Δ obtained from Eq.(4.16) is close to the smallest value possible, since Σ_t contains the smallest diagonal entries of Σ. Therefore for the system being in the balanced representation, the near-optimal reduction is obtained if the following condition is satisfied

$$\|\Sigma_t^2\| \ll \|\Sigma_r^2\| \tag{4.18}$$

This condition is often modified if the truncated parts B_t and C_t of B and C are not much larger than the retained parts B_r and C_r of B and C, i.e., if $\|C_t\| \gg \|C_r\|$, and $\|B_t\| \gg \|B_r\|$ do not hold. Since this is a very unusual situation, and rather degenerated case, therefore if in the balanced representation the diagonal entries of the Hankel matrix Γ are in a decreasing order, i.e., $\gamma_i \geq \gamma_{i+1} > 0$, $i=1,\dots,n-1$, the near-optimal reduction is obtained if

$$\|\Gamma_t^2\| \ll \|\Gamma_r^2\| \tag{4.19}$$

This condition says that the reduction is near-optimal if the sum of the Hankel singular values of the truncated states is much smaller than the sum of the Hankel singular values of the retained states.

The question arises as to whether the reduced model (A_r, B_r, C_r) is stable. The results of Moore [89], Parnebo and Silverman [94], and Glover [48] confirm it. They say that if a triple (A,B,C) is balanced, and the balanced grammians $\Gamma = diag(\gamma_i)$, $i = 1,\dots,n$, have their entries γ_i in decreasing order, $\gamma_i \geq \gamma_{i+1} \geq 0$, and if Γ is partitioned such that $\Gamma = diag(\Gamma_r, \Gamma_t)$, where Γ_r is positive definite, and if the triple (A,B,C) is partitioned accordingly, then (A_r, B_r, C_r) is stable.

For a system with complex poles, such as flexible structures, each pair of coordinates, rather than each single coordinate, is evaluated for reduction. This is a straightforward task, since the Hankel singular values related to each 2×2 block (i.e., related to two complex conjugate poles) are almost equal ($\gamma_i \cong \gamma_{i+1}$).

In the following examples the reduction in the balanced and modal coordinates is presented, using the Hankel singular values and the component costs. In the vast majority of cases, the reduction results are the same in both coordinates, and the same when using Hankel singular values and the costs. However, in the following examples we expose those not frequent and peculiar cases, where these coordinates give different results.

50

Example 4.1 A simple system with the masses $m_1=11$, $m_2=5$, $m_3=10$, stiffness $k_1=k_4=10$, $k_2=50$, $k_3=55$, and damping proportional to the stiffness, $d_i=0.01k_i$, $i=1,2,3,4$, is considered. The single input u is applied, such that $f_1=u$, $f_2=2u$, $f_3=-5u$, the output is a linear combination of the mass displacements, $y=2q_1-2q_2+3q_3$, where q_i is the displacement of the ith mass, and f_i is the force applied to that mass.

For this system the state space triple (A,B,C) is as follows

$$
A=\begin{bmatrix}
0 & 0 & 0 & 1 & 0 & 0 \\
0 & 0 & 0 & 0 & 1 & 0 \\
0 & 0 & 0 & 0 & 0 & 1 \\
-5.4545 & 4.5454 & 0 & -0.0545 & 0.0455 & 0 \\
10 & -21 & 11 & 0.1000 & -0.2100 & 0.1100 \\
0 & 5.5000 & -6.5000 & 0 & 0.0550 & -0.0650
\end{bmatrix}
$$

$B^{\mathrm{T}}=[0\ 0\ 0\ 0.0909\ 0.4\ -0.5]$, and $C=[2\ -2\ 3\ 0\ 0\ 0]$.

For this system in the balanced representation the reduction costs are as follows: $\sigma_1=1.9488$, $\sigma_2=1.9305$, $\sigma_3=0.4494$, $\sigma_4=0.4369$, $\sigma_5=0.3282$, and $\sigma_6=0.2932$, and the Hankel singular values of the same states are $\gamma_1=5.6089$, $\gamma_2=5.5967$, $\gamma_3=1.0998$, $\gamma_4=1.1139$, $\gamma_5=1.3447$, and $\gamma_6=1.3365$. In the case when the costs are in the decreasing order, the Hankel singular values are not ordered decreasingly. The reduction by deleting the two state variables corresponding to the two smallest reduction costs (σ_5 and σ_6) gives the reduction index of $\delta=0.1544$. The optimality index in this case is $\varepsilon=0.0001$, and the estimated index is $\delta_e=0.1545$. The reduction of the system by deleting the two state variables corresponding to the smallest two Hankel singular values (γ_3 and γ_4) gives a reduction index of $\delta=0.2208$, with the optimality index $\varepsilon=0.0008$ and the estimated index $\delta_e=0.2201$. The indices δ and ε show that the first solution is closer to the optimal one. Table 4.1 shows the reduction index δ and the optimality index ε versus k, the number of state variables of the reduced model. Observe the comparatively large values of ε for the odd k. This can be explained by the violation of the property (4.8) of the balanced representation and consequently the condition for the near-optimal reduction (4.19). Namely, each mode has a pair of almost equal Hankel singular values. For odd k the Hankel singular values of a pair are separated, and in this case the condition (4.19) is violated.

Table 4.1. Reduction and optimality indices[*]

k	δ	ε
1	1.207	0.560
2	0.269	0.0000007
3	0.319	0.09
4	0.154	0.0001
5	0.186	0.065

[*]from Ref.[38], courtesy of the publisher

4.3 Near-Optimal Reduction in Modal Coordinates

In the modal coordinates the matrix A is diagonal, in the representation (A,B,C) as in (4.1), thus, $A_{rt}=A_{tr}^T=0$. Since $y=y_r+C_t x_t$, the error $y\text{-}y_r$ between the original and the reduced model is solely determined from the parameters of the truncated system, i.e., $y\text{-}y_r=C_t x_t$. Therefore

$$\Delta^2=\|y\text{-}y_r\|^2=tr(C_t^T C_t W_{ct}) \tag{4.20a}$$

Similarly, when the observability grammian is considered, the norm is

$$\Delta^2=tr(B_t B_t^T W_{ot}) \tag{4.20b}$$

or

$$\Delta^2=-tr((A_t+A_t^T)W_{ct}W_{ot}) \tag{4.20c}$$

Note that for flexible structures grammians in modal coordinates are diagonally dominant, that is, one obtains $W_c W_o \cong \Gamma^4$. Thus the conditions for near-optimal reduction in modal coordinates are the same as in the balanced coordinates, i.e., as in Eqs.(4.18) and (4.19). However, the reduction in the modal coordinates does not shift the poles.

For A in modal form 2 one obtains $A_i + A_i^T = -2\zeta_i\omega_i I_2$, and from Eq.(3.60) it follows that $\gamma_i^2 \cong c_i^T c_i / 2\zeta_i\omega_i$. Hence, each modal cost evaluated from Eq.(4.20), yields to

$$\sigma_i^2 = \Delta_i^2 = \frac{\beta_{ii}\theta_{ii}}{2\zeta_i\omega_i} \tag{4.21}$$

which is the modal cost obtained by Skelton and Yousuff, Ref.[103].

Example 4.2 A reduction using component cost in modal coordinates for the system from Example 4.1 is presented. The reduced model has four state variables, and its state-space representation is as follows

$$A_r = \begin{bmatrix} -0.0191 & 0.8738 & 0 & 0 \\ -0.8738 & -0.0191 & 0 & 0 \\ 0 & 0 & -0.06562 & 5.1230 \\ 0 & 0 & -5.1230 & -0.0656 \end{bmatrix}, \quad B_r = \begin{bmatrix} -0.00001 \\ 0.1081 \\ -0.0048 \\ 0.3085 \end{bmatrix},$$

$$C_r = [-2.2182 \quad -0.00018 \quad -1.0427 \quad -0.01313]$$

which gives the optimality index $\varepsilon = 0.0002$, the normalized error $\delta = 0.1538$, and its estimation $\delta_e = 0.1539$. Comparing this example with Example 4.1, it is seen that in both cases the solutions are close to the optimal one, and that the modal reduced model is about as close to the optimal model as the balanced one.

Finally, the near-optimal results are compared with the optimal results of Wilson [117] and Hyland-Bernstein [55]. In Table 4.2 and 4.3 the results from Example 6.1 and 6.2 of Wilson are compared with the near-optimal balanced results. Both examples show that the reduced model obtained in modal or balanced coordinates is close to the optimal one obtained by Hyland and Bernstein. Note, however, that the optimal

53

solution in Example 6.2 from Hyland and Bernstein needs 14 iterations, while the near-optimal results are obtained in "a single shot."

Table 4.2. Reduction and optimality indices[*]

k	δ [55]	δ [116]	δ (bal)	ε [116]	ε (bal)
1	0.4268	---	0.4321	---	0.0159
2	0.0393	0.0410	0.0294	0.0032	0.0006
3	0.0013	---	0.0013	---	0.0001

[*]from Ref.[38], courtesy of the publisher

Table 4.3. Reduction and optimality indices[*]

	δ	ε
Hyland [55]	0.0975329	---
Balanced	0.0975333	0.0020
Modal	0.0975333	0.0020

[*]from Ref.[38], courtesy of the publisher

4.4 Closed-Form Near-Optimality Conditions.

In this section the near-optimal conditions for special but important cases are derived in a closed-form. For this purpose define the controllability and observability coefficients

$$\rho_{cij}^2 = \frac{\|W_{cij}\|^2}{\|W_{cii}\|\|W_{cjj}\|}, \qquad \rho_{oij}^2 = \frac{\|W_{oij}\|^2}{\|W_{oii}\|\|W_{ojj}\|} \qquad (4.22)$$

then one can see that the grammians are diagonally dominant if $\rho_{cij} \ll 1$ and $\rho_{oij} \ll 1$ for $i \neq j$. Thus for the near-optimal reduction it is sufficient that that $\rho_{cij} \ll 1$ and $\rho_{oij} \ll 1$ $i \neq j$, where i denotes a retained mode, and j denotes a deleted mode.

It follows from Eq.(3.69) that the ij-th block of the controllability grammian can be decomposed such that $W_{cij} = \beta_{ij} \tilde{W}_{cij}$, where $\tilde{W}_{cij}$ is purely a function of system properties (modal damping ζ_i

54

and ζ_j, and natural frequencies ω_i and ω_j), while β_{ij} purely depends on the actuator locations. Similarly, one can obtain for the observability grammian $W_{oij} = \theta_{ij} W_{oij}$, where again W_{oij} is purely a function of system properties, while θ_{ij} depends only on the sensor locations. Thus, the correlation coefficients can be also decomposed in the similar fashion, obtaining

$$\rho_{cij} = \kappa_{cij} r_{cij} \ll 1, \qquad \rho_{oij} = \kappa_{oij} r_{oij} \ll 1 \qquad (4.23a)$$

where

$$\kappa_{cij} = \kappa_{ij}, \qquad \kappa_{ij} = \frac{\|c_i^T c_j\|}{\|c_i\| \|c_j\|}, \qquad r_{cij}^2 = \frac{\|W_{cij}\|^2}{\|W_{cii}\| \|W_{cjj}\|}, \qquad (4.23b)$$

$$r_{oij}^2 = \frac{\|W_{oij}\|^2}{\|W_{oii}\| \|W_{ojj}\|}, \qquad W_{cij} = \frac{W_{cij}}{\beta_{ij}}, \qquad W_{oij} = \frac{W_{oij}}{\theta_{ij}}. \qquad (4.23c)$$

For a system with rate measurements only the controllability and observability grammians are identical up to the signs of its off-diagonal terms, thus the condition (4.23a) is now

$$\rho_{ij} = \kappa_{ij} r_{ij} \ll 1, \qquad (4.24a)$$

where

$$\kappa_{cij} = \kappa_{oij} = \kappa_{ij}, \qquad \kappa_{ij} = \frac{\|c_i^T c_j\|}{\|c_i\| \|c_j\|}, \qquad (4.24b)$$

and

$$r_{ij}^2 = \frac{\|W_{ij}\|^2}{\|W_{ii}\| \|W_{jj}\|}, \qquad W_{ij} = \frac{W_{cij}}{\beta_{ij}} = \frac{W_{oij}}{\theta_{ij}}. \qquad (4.24c)$$

Form the product of the controllability and observability

55

grammians, $W = W_c W_o$. If each block W_{ij} corresponding to a retained mode (i) and deleted mode (j), satisfies

$$\rho_{ij}^2 = \frac{\| W_{ij} \|^2}{\| W_{ii} \| \| W_{jj} \|} \ll 1 \tag{4.25}$$

then $\delta \ll 1$. Now, from (3.69), $W_{ij} = \beta_{ij} W_{ij}$ for W_{cij} is purely a function of ω_i, ω_j, ζ_i, and ζ_j, so the correlation coefficient ρ_{ij} can be written as a product

$$\rho_{ij} = \kappa_{ij} \eta_{ij} \tag{4.26}$$

where, from the Schwartz inequality, $\kappa_{ij}^2 = \beta_{ij}^2 / (\beta_{ii} \beta_{jj}) \leq 1$, and $\eta_{ij}^2 = \| W_{ij} \|^2 / (\| W_{ii} \| \| W_{jj} \|)$. The condition $\rho_{ij} \ll 1$ is now as follows

$$\kappa_{ij} \eta_{ij} \ll 1 \tag{4.27}$$

where η_{ij} depends purely on the structural properties, such as ζ_i, ζ_j, ω_i, and ω_j, rather than on sensor/actuator locations. This reflects the relationship between structural parameters and the optimality of reduction. Since $0 \leq \kappa_{ij} \leq 1$, the condition $\eta_{ij} \ll 1$ certainly implies $\rho_{ij} \ll 1$, but not vice versa.

In contrast, the coefficient κ_{ij} depends only on sensor/actuator locations. It can be seen that a near-optimal reduction will occur if the ith and jth rows of B (columns of C) are nearly orthogonal for all retained modes i and deleted modes j. Thus, condition (4.27) decouples in a simple way the contributions toward the optimality of reduction of the structural properties (η_{ij}) and sensor/actuator locations (κ_{ij}).

We now study the influence of the structural parameters ζ_i and ω_i on the optimality index. Introducing dimensionless variables $\omega = \omega_i / \omega_j$, and $\gamma = \zeta_i / \zeta_j$, one obtains

$$n_{ij}^2 = \frac{n}{m^2} \qquad\qquad (4.28a)$$

where

$$m = 4\zeta_j^2 \omega(\omega + \chi)(1 + \omega\chi) + (1 - \omega^2)^2 \qquad\qquad (4.28b)$$

and

$$n = 8\zeta_j^2 \omega\chi\{4\zeta^2\omega^2[(\omega + \chi)^2 + (1 + \omega\chi)^2] + (1 - \omega^2)^2(1 + \omega^2)\} \qquad (4.28c)$$

Plots of n_{ij} are shown in Fig.4.1 for the case $\zeta_i = \zeta_j = \zeta$, i.e. $\chi = 1$. The curves show that the condition (4.27) is satisfied for small ζ and separated natural frequencies ($\omega \neq 1$). In fact, the smaller the damping and the more separated the natural frequencies of the retained and deleted modes, the closer the reduction is to the optimal one. The importance of these graphs is that they apply for any structure. Once its damping coefficients ζ_i and natural frequencies ω_i are known, the n_{ij} can simply be read off, giving the modal coupling information directly.

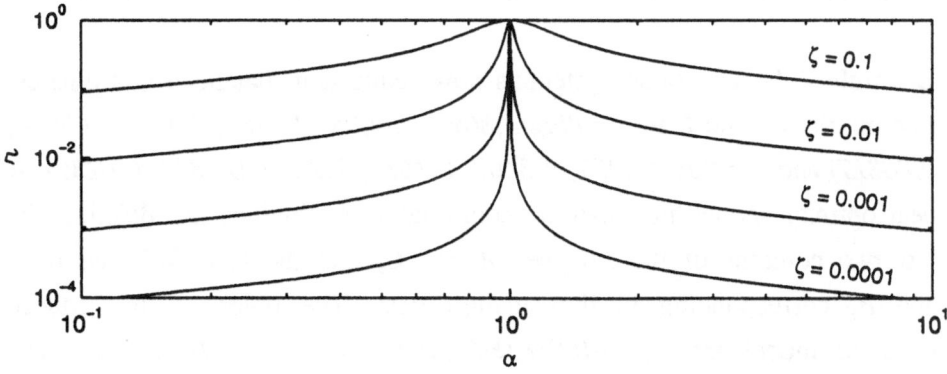

Figure 4.1: Correlation coefficient for rate only output.

Similar relationships were derived for the case of the rate measurements in Refs.[38] and [47], showing again that modes are weakly correlated for small damping and for separated poles.

Example 4.3 The flexible system from Fig.2.1 with two inputs and two outputs is investigated. The system parameters are as follows $m_1=m_2=m_3=10$, $k_1=k_4=1000$, $k_2=500$, $k_3=100$, $d_i=dk_i$, $i=1,...,4$, and

$$B^T=\begin{bmatrix} 0 & 0 & 0 & 0.1 & 0.2 & -0.5 \\ 0 & 0 & 0 & -0.2 & -0.3 & 0.1 \end{bmatrix}, \quad C=\begin{bmatrix} 1 & -1 & 1.5 & 0.5 & -2.5 & -1 \\ 0.5 & 1.5 & -0.5 & 1 & -1 & 0.5 \end{bmatrix}$$

A near-optimal reduced system with four states is desired, using modal truncation. The system poles for $d=0.005$ are $\lambda_{1,2}=-0.0182\pm j6.0492$, and $\lambda_{3,4}=-0.0554\pm j10.5291$, and $\lambda_{5,6}=-0.0863\pm j13.1352$, and they are separated. The system grammians are diagonally dominant, as expected, and the cost matrix $\Sigma=diag(5.5026, 3.8682, 4.8734, 0.1287, 0.0007, 0.1453)$ is obtained. Deleting the last two rows of A and B, and the last columns of A and C, the cost of which is the smallest, we obtain the near-optimal reduced model. The optimality index of the obtained model is small indeed, $\varepsilon= 0.0011$, and the reduction error is $\delta=0.0577$. The solid line in Fig.4.2 shows the dependence of the optimality index ε and the reduction error δ on the damping parameter d. As expected, the smaller the damping, the closer the reduced model is to optimal.

Example 4.4 The same system is now reduced in balanced coordinates. The matrices Γ and Σ are $\Gamma=diag(4.8097, 4.8095, 2.1781, 2.1765, 0.8084, 0.80817)$ and $\Sigma=diag(5.2837, 3.3607, 0.1618, 2.2239, 0.3832, 0.0239)$. A near-optimal model in balanced coordinates is obtained by deleting the last two columns of the balanced A and C, and the last two rows of A and B, corresponding to the smallest cost. The poles of the reduced balanced model are $\lambda_{r1,2}=-0.0183\pm j6.0492$, $\lambda_{r3,4}=-0.0554\pm j10.5291$. The obtained model is near-optimal, since its optimality index is $\varepsilon=0.6758\times10^{-5}$, and the reduction error is $\delta=0.0577$. Fig.4.2, dotted lines, shows the dependence of ε and δ on the damping parameter d. From the plot of ε it can be seen that, the smaller the damping is, the closer the solution is to the optimal one. For high damping the output error δ, by contrast, falls off, indicating that one of the balanced modes becomes nearly unobservable and unobservable. Note that the results obtained for low levels of damping are very close to those obtained in the previous example using modal coordinates. By contrast, for high damping the modal correlations increase to the point where balancing should be expected to give significantly better results than modal, which is confirmed by Fig.4.2.

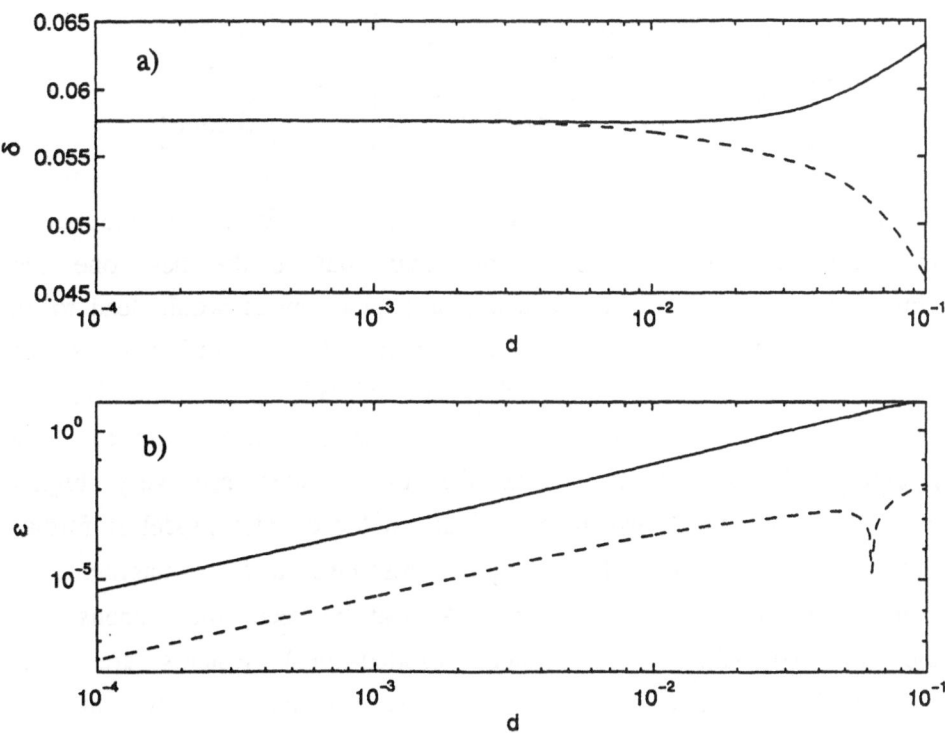

Figure 4.2: Effect of damping on reduction accuracy of the simple system: (a) reduction error, (b) optimality index

Balanced reduction gives a reduced-order model that is closer to optimal one than the modal reduction, if the structure considered has closely-spaced poles. The following examples confirms this fact.

Example 4.5 The system from Fig.2.1 is again considered, but now with the following parameters: $m_1 = m_3 = 1, m_2 = 2, k_1 = k_4 = 50, k_2 = k_3 = 5, d_i = 0.01k_i$, $i = 1,2,3,4$, and $B^T = [0\ 0\ 0\ 1\ 0\ 0]$, $C = [1\ 2\ 2\ 0\ 0\ 0]$.

This system is reduced from six to four states. The system poles are $\lambda_{1,2} = -0.0225 \pm j2.1224$, $\lambda_{3,4} = -0.2750 \pm j7.4111$, $\lambda_{5,6} = -0.2775 \pm j7.4443$. The second and the third pair of poles are clearly closely spaced. The reduced model using modal truncation is:

$$A_r = \begin{bmatrix} -0.0225 & 2.1224 & 0 & 0 \\ -2.1224 & -0.0225 & 0 & 0 \\ 0 & 0 & -0.2774 & 7.4443 \\ 0 & 0 & -7.4443 & -0.2774 \end{bmatrix}, \quad B_r = \begin{bmatrix} -0.0002 \\ 0.0495 \\ -0.1580 \\ 0.6808 \end{bmatrix}$$

and $C_r = [1.0719\ 0.0041\ 0.2598\ 0.0603]$, with $\varepsilon = 0.7050$ and $\delta = 0.3063$. In this case the result is not near-optimal, but is the best one can obtain in modal coordinates. Deleting any other mode would lead to an even larger reduction error. The source of the difficulty is, of course, the close second and third pair of poles. The controllability and observability coefficients for these two modes are $\rho_{c23} = \rho_{o23} = 0.9964$, showing that the two modes are very highly correlated. The poor performance of this reduced-order model is further illustrated by the plots of the impulse response, and its magnitude of transfer function, given in Fig.4.3. As can be seen, the responses of the full (solid-lines) and reduced (dotted-lines) systems are quite different. The same structure is now re-examined in balanced coordinates. The poles of the reduced system are $\lambda_{r1,2} = -0.0225 \pm j2.1224$ and $\lambda_{r3,4} = -0.2753 \pm j7.4630$, i.e., the two close poles have now been separated during the reduction. The optimality indices obtained are $\varepsilon = 0.54 \times 10^{-5}$ and $\delta = 0.0017$, so the reduction is near-optimal. Indeed, the plots of impulse response and magnitude of transfer function for this reduced system overlap those of the full system in Fig.4.3. This example, taken together with the preceding one, illustrates the expected result that the high correlation of close modes prevents near-optimal model reduction using modal truncation, but presents no such problems in balanced coordinates.

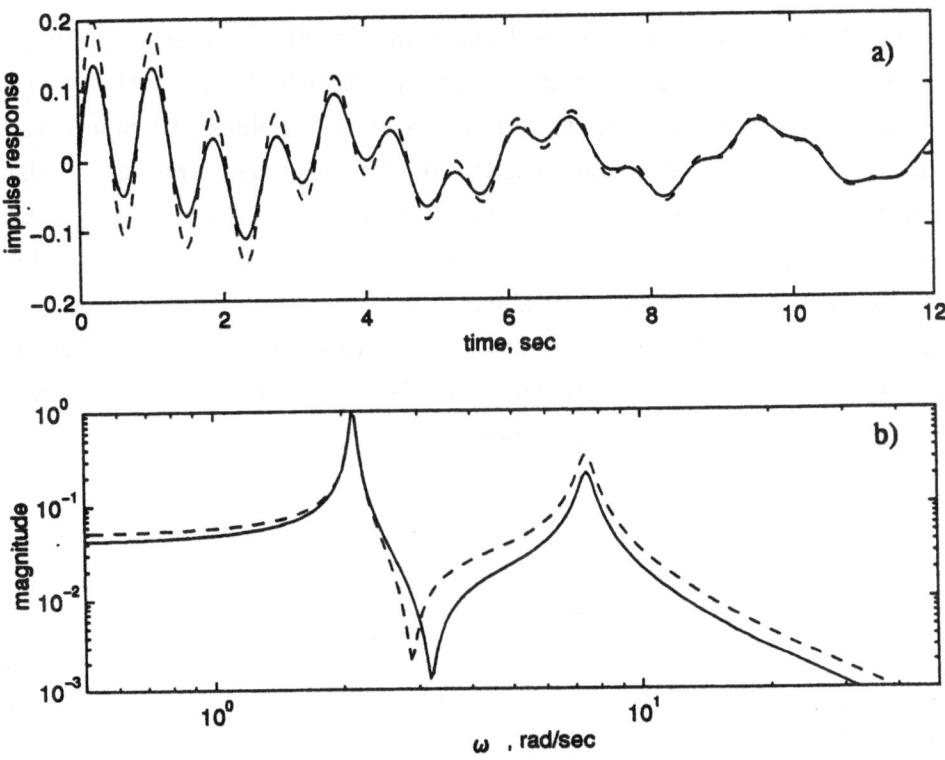

Figure 4.3: Full and reduced-order models (full model and reduced balanced model - solid line, reduced modal model - dashed line): (a) impulse responses, and (b) magnitudes of transfer function

Example 4.6 The reduction of a flexible truss model in Fig.2.2 is next presented. The truss damping is proportional to the stiffness matrix, $d=0.0002k$, the vertical input force is located at node $n7$, and the vertical displacement of node $n4$ is the system output. The truss model was reduced in balanced coordinates. The results are presented in Fig.4.4. It shows that the reduction from 26 to 6 states causes a reduction error, δ, of less than 10%. The plots of the impulse response and magnitude of transfer function are shown in Fig.4.5.

Example 4.7 The model of the Control of Flexible Structures (COFS-1) mast, Fig.4.6, consists of its structural properties as well as its actuator and sensor dynamics. It has 104 state variables, 10 inputs and 20 outputs. The reduction results, in balanced as well as modal coordinates, are presented in Fig.4.7 (balanced reduction in dotted line; modal reduction, solid line). They show that reduction from 104 to 38 states causes an error of about 10%, and to 60 states an error of about 2%, in both balanced and modal coordinates. Plots in Fig.4.8 show the magnitude of transfer function of the full and reduced (60-state) model, with input No. 8 and output Nos. 17 and 18.

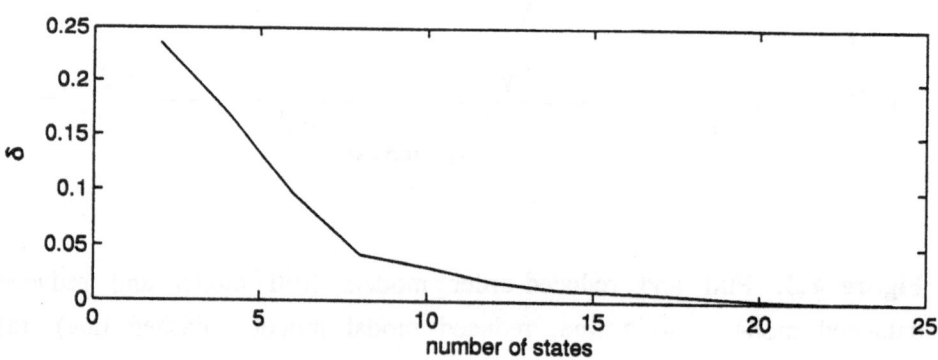

Figure 4.4: Truss reduction error vs. order of the reduced model

4.5 Reduction in Finite Time and Frequency Intervals

Having defined and analyzed time-limited and band-limited grammians, one can use them to perform reduction of a system order such that the dynamics of full- and reduced-order models are close to each other within the finite time sequence or finite frequency band. This approach is useful, for example, in the model reduction of unstable plants (using time-limited grammians), or in the filter design (using band-limited grammians).

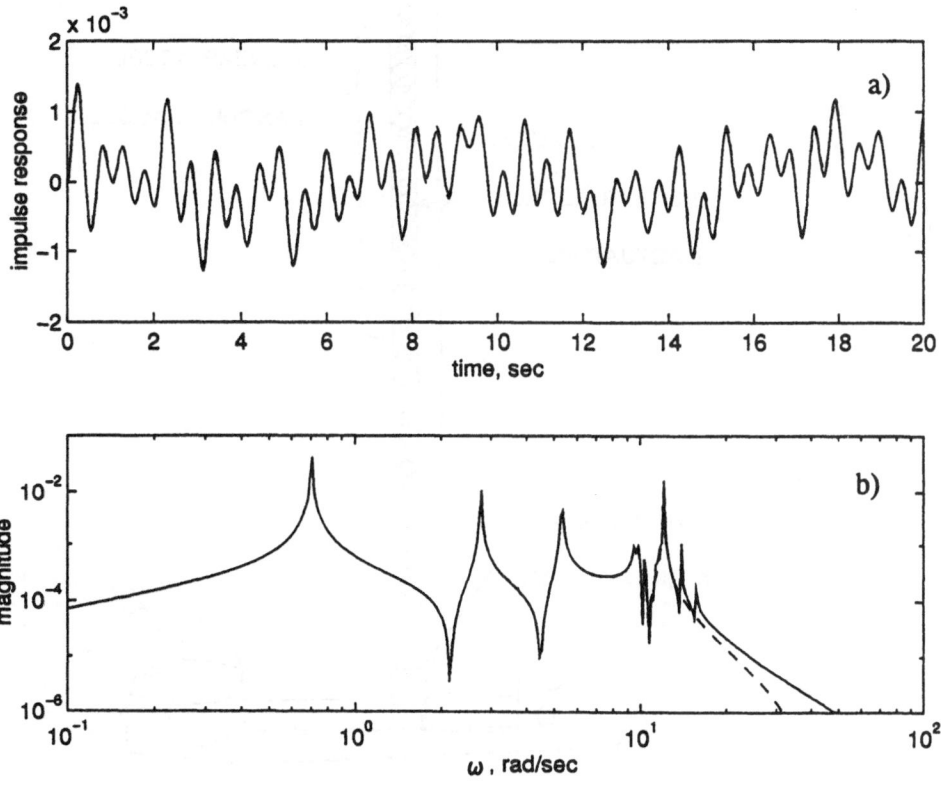

Figure 4.5: Full (solid line) and reduced-order (dashed line) models of the truss: (a) impulse responses, and (b) magnitudes of transfer function

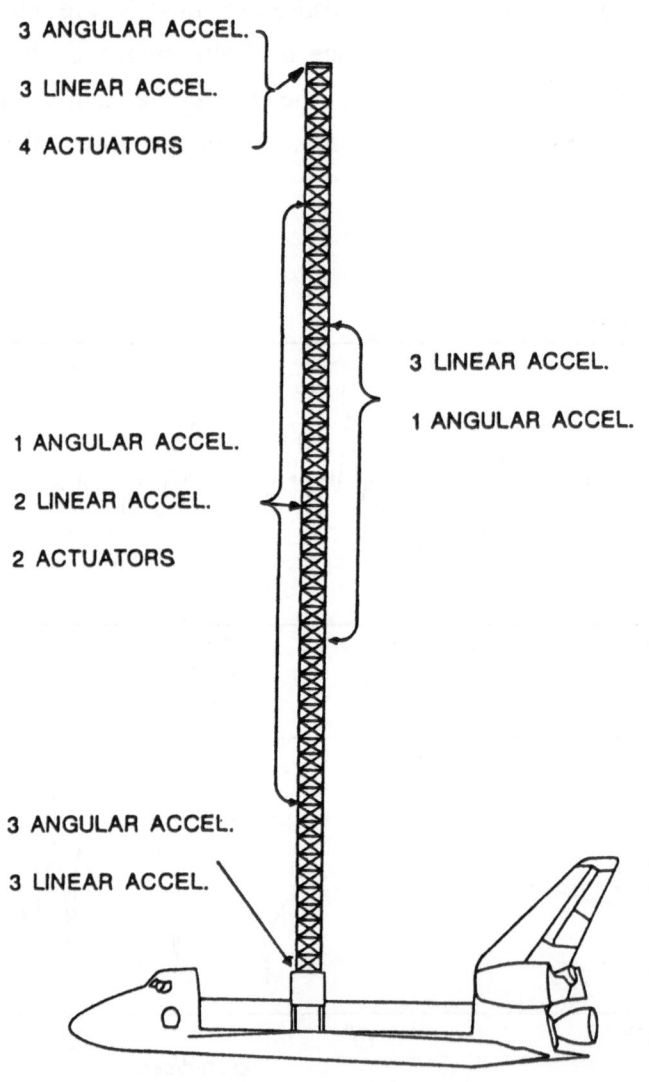

3 ANGULAR ACCEL.

3 LINEAR ACCEL.

4 ACTUATORS

3 LINEAR ACCEL.

1 ANGULAR ACCEL.

1 ANGULAR ACCEL.

2 LINEAR ACCEL.

2 ACTUATORS

3 ANGULAR ACCEL.

3 LINEAR ACCEL.

Figure 4.6: COFS-1 structure (from Ref.[38], courtesy of the publisher)

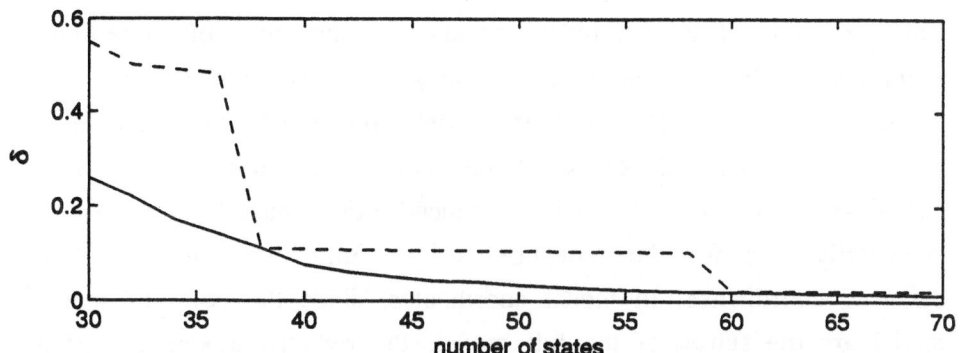

Figure 4.7: COFS-1 reduction error vs. order of the reduced model

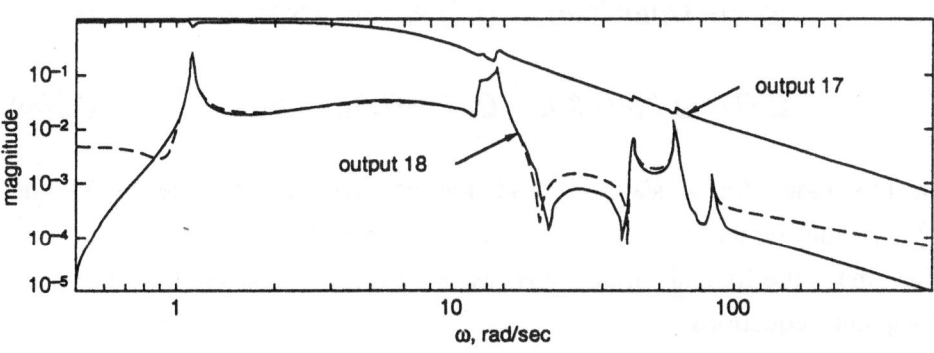

Figure 4.8: Magnitudes of transfer function of full and reduced-order models of the COFS-1

4.5.1 Reduction in Finite Time Interval

In the finite time interval, the reduction is similar to the infinite time interval. The difference occurs in the use of time-limited grammians instead of the infinite time grammians, and in the stability conditions for the reduced-order model. While the reduction in the infinite time interval of a stable system in modal or balanced coordinates produces a stable reduced-order model, this is not necessarily true for the reduction in the finite time interval in the balanced coordinates. In modal coordinates, since the poles of reduced model are the subset of the full model, the reduced model of a stable system is stable. A different situation is observed in the balanced coordinates. In order to establish stability conditions for the reduced balanced model of a finite time interval, assume the original system to be stable, and for $S(t)$ as in (3.9) denote the auxiliary matrices $Q_c(T)$, $Q_o(T)$

$$Q_c(t) = S(t)BB^T S^T(t), \quad Q_o(t) = S^T(t)C^T CS(t) \qquad (4.29a)$$

$$Q_c(T) = Q_c(t_1) - Q_c(t_2), \quad Q_o(T) = Q_o(t_1) - Q_o(t_2) \qquad (4.29b)$$

In this case, for a stable A and for positive semidefinite $Q_c(T)$ and $Q_o(T)$, the reduced balanced model is stable. To prove it, note first that the time-limited grammians as in Eq.(3.8) satisfy the following Lyapunov equations

$$AW_c(T) + W_c(T)A^T + Q_c(T) = 0, \quad A^T W_o(T) + W_o(T)A + Q_o(T) = 0 \qquad (4.30)$$

If the matrices $Q_c(T)$ and $Q_o(T)$ are positive semidefinite, then according to the stability theorem of Moore, [89] Parnebo and Silverman [94] and Glover [48], the reduced balanced model is stable.

Denote $\Delta t = t_2 - t_1$, and

$$Q_c(\Delta t)=BB^T-S(\Delta t)BB^T S^T(\Delta t), \qquad Q_o(\Delta t)=C^T C-S^T(\Delta t)C^T CS(\Delta t) \qquad (4.31a)$$

With this notation one obtains $Q_c(T)=S(t_1)Q_c(\Delta t)S^T(t_1)$, $Q_o(T)=S^T(t_1)Q_o(\Delta t)S(t_1)$, from which it follows that $Q_c(T)$ and $Q_o(T)$ are positive semi-definite if $Q_c(\Delta t)$ and $Q_o(\Delta t)$ are positive semi-definite. But, for stable systems

$$\lim_{\Delta t\to\infty} S(\Delta t) \ = \ \lim_{\Delta t\to\infty} exp(A\Delta t)=0 \qquad (4.31b)$$

thus $\lim\limits_{\Delta t\to\infty} S(\Delta t)BB^T S^T(\Delta t)=0$ and $\lim\limits_{\Delta t\to\infty} S^T(\Delta t)C^T CS(\Delta t)=0$, and $\lim\limits_{\Delta t\to\infty}$ $Q_c(\Delta t)=BB^T$, $\lim\limits_{\Delta t\to\infty} Q_o(\Delta t)=C^T C$. Consequently, the matrices $Q_c(\Delta t)$ and $Q_o(\Delta t)$ are positive-definite, and the reduced balanced model in the finite time interval is stable if the time interval Δt is large enough. If the time interval is too short to obtain a stable balanced reduction, and a stable model is required, then either the interval should be enlarged, or modal reduction applied. The last reduction gives always a stable model for a stable full-order system.

The computational procedure is summarized as follows:
1. For given (A,B,C) determine grammians W_c, W_o from (3.3).
2. Compute $W_c(t_i)$, $W_o(t_i)$, and $S(t_i)$, $i=1,2$, from (3.8) and (3.9).
3. Determine $W_c(T)$ and $W_o(T)$ from (3.7).
4. Apply the reduction procedure to obtain (A_r,B_r,C_r) for $W_c(T)$ and $W_o(T)$.
5. Compute δ and ε for the reduced model (A_r,B_r,C_r) from (4.5) and (4.10).

The following examples illustrate the application of the time-limited grammians in the model reduction. In these examples the variables in modal coordinates are denoted with the subscript "m," while those in balanced coordinates are indicated by the subscript "b."

67

The resulting plots present the full system response as well as the reduced model responses. The full system response is plotted with a solid line, the reduced balanced response with a dotted line, and the reduced modal response with a dashed line.

Example 4.8 A simple system from the Example 4.1 is considered. The system poles are $\lambda_{1,2}=-0.0038\pm j0.8738$, $\lambda_{3,4}=-0.0297\pm j2.4374$, and $\lambda_{5,6}=-0.1313\pm j5.1217$. Two cases of model reduction from 6 to 4 state variables are considered: Case 1, the reduction over the interval $T_1=[0, 8]$, and Case 2, the reduction over the interval $T_2=[10, 18]$.

In Case 1, the first and the third pair of poles are preserved in the reduced modal model, while in the balanced coordinates the poles of the reduced model are $\lambda_{b1,2}=-0.0009\pm j0.8712$, and $\lambda_{b3,4}=-0.1395\pm j5.0890$. The reduction error is $\delta_b=0.2954$, and $\delta_m=0.2911$, and the optimality index is $\varepsilon_b=0.0006$, and $\varepsilon_m=0.0129$. Since the optimality indices are much smaller than 1, and thus the reduced models are near-optimal. The impulse responses for the balanced and modal reductions are compared in Fig.4.9a, where the area outside the interval T_1 is shaded.

Case 2 is obtained from Case 1 by shifting the interval T_1 by 10. In this case, the first two pair of poles are preserved in modal reduction, and $\lambda_{b1,2}=-0.0039\pm j0.8742$, $\lambda_{b3,4}=-0.03047\pm j2.4439$ are the poles of the reduced balanced model. The reduction error in this case is $\delta_b=0.2919$, and $\delta_m=0.3012$, and the reduced model is near-optimal, since $\varepsilon_b=0.0625$, $\delta_m=0.0490$. The impulse responses of the reduced and original systems are presented in Fig.4.9b for Case 2 where, again, the area outside the time interval T_2 is shaded. Comparison of Figs.4.9a and 4.9b shows that the third mode is less visible for $t>10$ sec, thus this mode was eliminated.

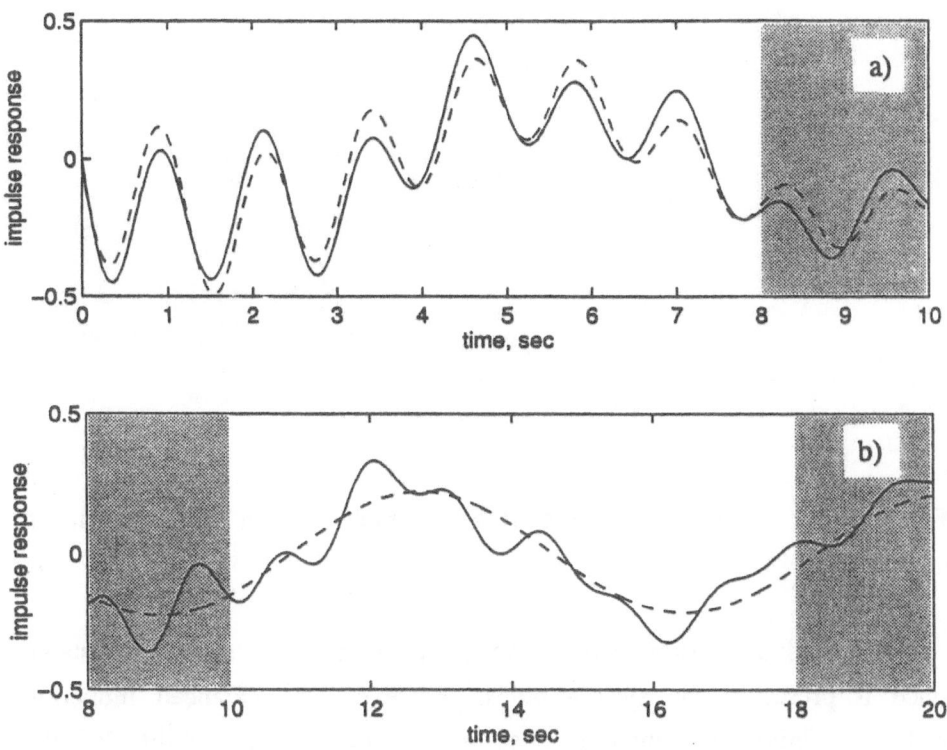

Figure 4.9: Impulse responses of a simple system and its reduced models for: (a) $T=[0, 8]$ sec, and (b) $T=[10, 18]$ sec

Example 4.9 Example 4.8 is re-examined, with the negative damping $d_i=-0.006k_i$, $i=1,2,3,4$, which makes the system unstable. The system poles are $\lambda_{1,2}=0.0023\pm j0.8738$, $\lambda_{3,4}=0.0178\pm j2.4375$, and $\lambda_{5,6}=0.0787\pm j5.1228$. The system grammians are determined over the interval $[0, 8]$ sec. The reduced balanced model of order 4 has the following poles $\lambda_{b1,2}=0.0059\pm j0.8716$, $\lambda_{b3,4}=0.0748\pm j5.1238$, whereas the first and the third pairs of poles are preserved in the reduced modal model. The reduction error is $\delta_b=0.2117$ and $\delta_m=0.2122$, and optimality index is $\varepsilon_b=0.0320$ and $\varepsilon_m=0.0200$. The reduced models are obviously near-optimal.

69

The impulse responses of the original and reduced models, within the interval [0, 8] sec, are shown in the nonshaded area of Fig.4.10.

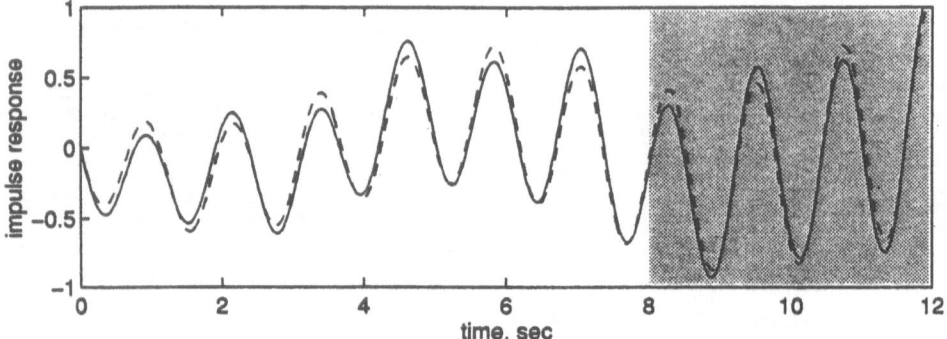

Figure 4.10: Impulse responses of an unstable system and its reduced models

The reduced modal model, obtained for a stable system, is stable since it preserves the system poles. However, the balanced reduction within a finite time interval does not guarantee the stability of the reduced model. This can be heuristically explained by the fact that within a finite time interval, especially within a short interval, the response of an unstable system can approximate the response of a stable system (and *vice versa*) with an acceptable error. The following example illustrates the case.

Example 4.10 Example 4.8 is re-examined with damping $d=0.006k$, so that the system is stable. The reduced balanced model obtained for the time interval $T=[0, 8]$, has the following poles: $\lambda_{1,2}=0.0009\pm j0.8713$, and $\lambda_{3,4}=-0.0857\pm j5.1127$, i.e., it is unstable. The reduced modal model is stable, with poles $\lambda_{m1,2}=-0.0023\pm j0.8738$, and $\lambda_{m3,4}=-0.0788\pm j5.1228$. The reduction error is $\delta_b=0.2807$ and $\delta_m=0.2810$, and the optimality index is $\varepsilon_b=0.0119$ and $\delta_m=0.0086$. The impulse responses of the original and reduced models are shown in Fig.4.11, where the area outside the interval T is shaded. The matrices $Q_c(T)$, $Q_o(T)$ for the reduced balanced model are not positive semi-definite, since their eigenvalues are $\lambda_{Qct} = \{0.4168, -0.1999, 0, 0, 0, 0\}$, $\lambda_{Qot} = \{16.6741, -4.7071, 0, 0, 0, 0\}$, which indicate an unstable reduced balanced model.

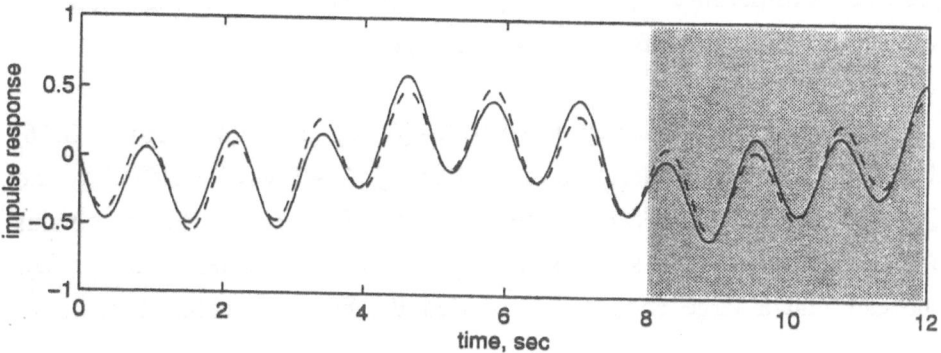

Figure 4.11: Impulse responses of a simple system and its reduced models

4.5.2 Reduction in Finite Frequency Interval

One approach to reduce a model in frequency domain is to impose frequency weighting on input and/or output, see Refs.[26], [119], and [14]. Model reduction in a finite frequency interval uses the band-limited grammians rather than weighting, and this approach is formally the same as the standard problem presented earlier. A balanced representation and Hankel singular values are determined from the band-limited grammians, and the states with small singular values are truncated. The question arises, as to whether a reduced system obtained this way is stable.

The stability condition for the balanced reduction in a finite frequency interval says that for a stable A and positive semidefinite $Q_c(\Omega)$ and $Q_o(\Omega)$, the reduced balanced model is stable (the matrices $Q_c(\Omega)$ and $Q_o(\Omega)$ are defined in Eqs.(3.24) and (3.26), respectively).

In order to prove it, note that the band-limited grammians $W_c(\Omega)$ and $W_o(\Omega)$ satisfy the Lyapunov equations (3.23) and (3.25). And for positive semi-definite $Q_c(\Omega)$ and $Q_o(\Omega)$, the reduced balanced model is stable, according to Refs.[89], [94], and [48].

Consider the frequency interval $\Omega=[0, \omega]$. It follows from (3.16) that $S(\omega)=0$ for $\omega=0$, therefore one obtains $Q_c(\Omega)=S(\omega)BB^T+BB^TS^*(\omega)$ and $Q_o(\Omega)=S^*(\omega)C^TC+C^TCS(\omega)$. At the same time, from Eq.(3.16) one obtains $S(\omega)=1/2$ for $\omega \to \infty$. As a result, for the frequency interval $\Omega=[0, \omega]$, one can find a large enough ω such that the balanced reduced model is stable.

The computational procedure is summarized as follows:
1. Determine stationary grammians W_c and W_o from (3.3) for a given (A,B,C).
2. Determine $W_c(\omega_1)$ and $W_c(\omega_2)$ from (3.15), and $W_o(\omega_1)$ and $W_o(\omega_2)$ from Eq.(3.17).
3. Determine $W_c(\Omega)$ and $W_o(\Omega)$ from (3.18).
4. Apply the reduction procedure to obtain the reduced state-space triple (A_r,B_r,C_r) using grammians $W_c(\Omega)$, $W_o(\Omega)$.
5. Determine indexes δ, and ε from (4.5), and (4.10) for the reduced model (A_r,B_r,C_r).

Example 4.11 Example 4.8 is considered in two cases: Case 1, in the frequency interval $\Omega_1=[0.7, 3.2]$, and Case 2, in the frequency interval $\Omega_2=[1.5, 3.2]$. In the first case, the reduced model with four state variables is determined, whereas in the second case, the reduced model with two state variables is obtained. In Case 1 the reduction errors are $\delta_b=0.0341$ and $\delta_m=0.0353$, and the optimality indices are $\varepsilon_b=0.0028$ and $\varepsilon_m=0.0234$, while in Case 2 one obtains $\delta_b= 0.1086$, $\delta_m=0.1331$, $\varepsilon_b=0.0109$, and $\varepsilon_m=0.1145$. In either case, the reduced model obtained is close to the optimal one ($\varepsilon_b \ll 1$, and $\varepsilon_m \ll 1$). Figures 4.12a and 4.12b (nonshaded areas) show a good fit of the magnitudes of the transfer function within the frequency interval considered, indicating that the reduction in the finite frequency interval can serve as a filter design method. For example, in Case 2 the output signal is filtered such that the resulting output of the reduced model is best fitted to the output of the original system within the interval $\Omega_2 = [1.5, 3.2]$ rad/sec.

4.5.3 Reduction in Finite Time and Frequency Intervals

The reduction technique is similar to that above, however the question of the stability of the reduced-order model has to be solved. To determine stability conditions for the reduced model in finite time and frequency intervals, note that $W_c(T,\Omega)$ and $W_o(T,\Omega)$ satisfy the following Lyapunov equations

$$W_c(T,\Omega)A^T+AW_c(T,\Omega)+Q_c(T,\Omega)=0, \tag{4.32a}$$

$$W_o(T,\Omega)A+A^TW_o(T,\Omega)+Q_o(T,\Omega)=0 \tag{4.32b}$$

where the matrix $Q_c(T,\Omega)$ is defined as follows

$$Q_c(T,\Omega)=Q_c(t_1,\Omega)-Q_c(t_2,\Omega), \quad Q_c(t,\Omega)=Q_c(t,\omega_2)Q_c(t,\omega_1) \tag{4.33a}$$

$$Q_c(t,\omega)=Q_c(t)S^*(\omega)+S(\omega)Q_c(t)=S(t)Q_c(\omega)S^*(t) \tag{4.33b}$$

and $Q_c(t)$ and $Q_c(\omega)$ are given by (4.29) and (3.21), respectively. Similarly is defined matrix $Q_o(T,\Omega)$

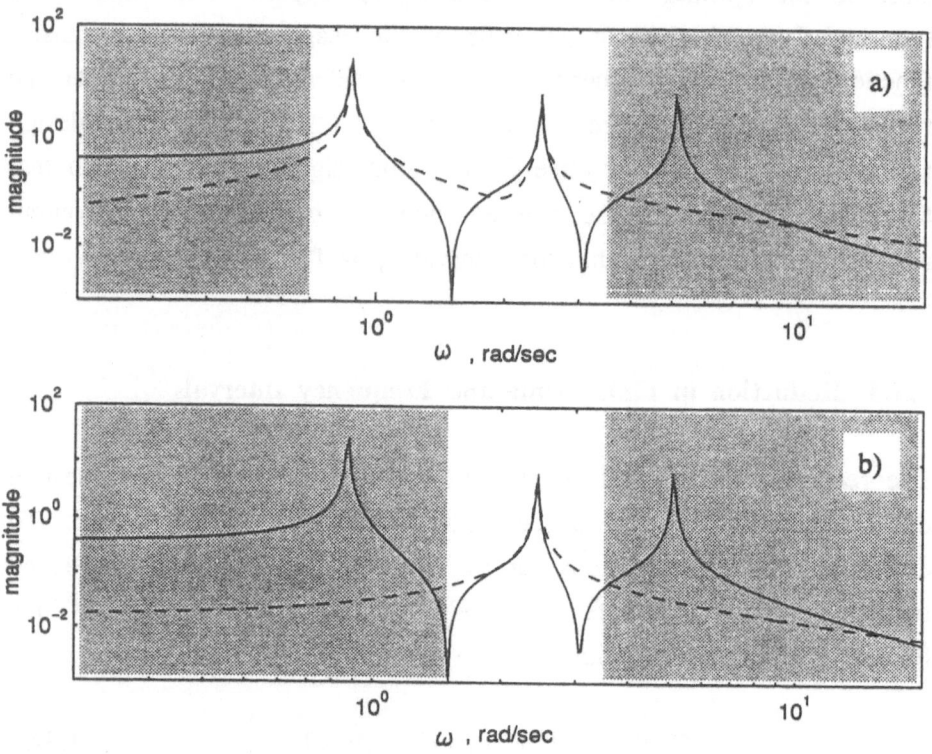

Figure 4.12: Magnitudes of transfer function of a simple system and its reduced models for: (a) $\Omega=[0.7,\ 3.2]$ rad/sec, and (b) $\Omega=[1.5,\ 3.2]$ rad/sec

$$Q_o(T,\Omega)=Q_o(t_1,\Omega)-Q_o(t_2,\Omega), \quad Q_o(t,\Omega)=Q_o(t,\omega_2)Q_o(t,\omega_1) \qquad (4.34a)$$

$$Q_o(t,\omega)=Q_o(t)S(\omega)+S^*(\omega)Q_o(t)=S^*(t)Q_o(\omega)S(t) \qquad (4.34b)$$

and $Q_o(t)$ and $Q_o(\omega)$ are given by (4.29) and (3.21) respectively.

A condition for the stability of the reduced-order model in finite time and frequency interval says that for a stable A and positive semi-definite $Q_c(T,\Omega)$ and $Q_o(T,\Omega)$, the reduced balanced model is stable. For positive semi-definite $Q_c(T,\Omega)$ and $Q_o(T,\Omega)$ the proof is similar to the finite-time case. Again, large time and frequency intervals result in a stable reduced balanced model.

The computational procedure can be set up alternatively, either by first applying frequency and then time transformation of grammians, or by first applying time and then frequency transformation. Since both procedures are similar, only the first one is presented as follows:

1. Determine grammians W_c and W_o, from (3.3) for given (A,B,C).
2. Determine $W_c(\omega_1)$ and $W_c(\omega_2)$ from (3.15), and $W_o(\omega_1)$ and $W_o(\omega_2)$ from (3.17).
3. Determine $W_c(t_i,\omega_j)$ and $W_o(t_i,\omega_j)$ from (3.35) and (3.38), for $i,j=1,2$.
4. Determine $W_c(t_i,\Omega)$ and $W_o(t_i,\Omega)$ from (3.34) and (3.37), for $i=1,2$.
5. Determine $W_c(T,\Omega)$ and $W_o(T,\Omega)$ from (3.33) and (3.36).
6. Apply the reduction procedure to obtain (A_r,B_r,C_r) using grammians $W_c(T,\Omega)$ and $W_o(T,\Omega)$.
7. Determine δ, and ε from (4.5), and (4.10) for the reduced model (A_r,B_r,C_r).

Matlab function *bal_time_freq* given in Appendix B3 determines the balanced representation in either time, or frequency, or both.

Example 4.12 Example 4.8 is considered in the time interval $T=[0,\ 8]$ and the frequency interval $\Omega=[0.7,\ 3.2]$. The obtained reduction errors are $\delta_b=0.1153$ and $\delta_m=0.1358$, and the optimality indices are $\varepsilon_b=0.0406$ and $\varepsilon_m=0.0260$. Figures 4.13a,b (nonshaded areas) show the impulse responses and the magnitudes of transfer function of the original system and the reduced balanced model within the intervals T and Ω.

This example can be considered as a combination of Examples 4.8 and 4.11, although the results are different. In Example 4.8 for the time interval $T=[0, 8]$ the second pole was deleted. In the present example, the third pole is deleted as a result of the additional restriction on the frequency interval.

4.5.4 Applications

Model reduction of the Advanced Supersonic Transport and the flexible truss structure are performed.

Advanced Supersonic Transport. The model of the Advance Supersonic Transport, was taken from Colgren [16]; it is a linear unstable system of order eight, with four inputs and eight outputs. In order to make our presentation concise, the model presented here is restricted to one output. The system triple (A,B,C) is

$$A=\begin{bmatrix} -0.0127 & -0.0136 & -0.0360 & 0.0000 & 0.0000 & 0.0000 & 0.0000 & 0.0000 \\ -0.0969 & -0.4010 & 0.0000 & 0.9610 & 19.59 & -0.1185 & -9.2000 & -0.1326 \\ 0.0000 & 0.0000 & 0.0000 & 1.0000 & 0.0000 & 0.0000 & 0.0000 & 0.0000 \\ -0.2290 & 1.7260 & 0.0000 & -0.7220 & -12.021 & -0.3420 & 1.8422 & 0.8810 \\ 0.0000 & 0.0000 & 0.0000 & 0.0000 & 0.0000 & 1.0000 & 0.0000 & 0.0000 \\ 0.0000 & 0.1204 & 0.0000 & 0.0496 & -44.000 & -1.2741 & -4.0301 & -0.5080 \\ 0.0000 & 0.0000 & 0.0000 & 0.0000 & 0.0000 & 0.0000 & 0.0000 & 1.0000 \\ 0.0000 & 0.1473 & 0.0000 & 0.3010 & -7.4901 & -0.1257 & -21.700 & -0.8030 \end{bmatrix}$$

$$B^T=\begin{bmatrix} 0.0000 & -0.0215 & 0.0000 & -1.0970 & 0.0000 & -0.6400 & 0.0000 & -1.8820 \\ 0.0194 & 0.0000 & 0.0000 & 0.0000 & 0.0000 & 0.0000 & 0.0000 & 0.0000 \\ 0.0000 & -0.0040 & 0.0000 & 0.3660 & 0.0000 & 0.1625 & 0.0000 & 0.4720 \\ 0.0000 & -1.7860 & 0.0000 & -0.0569 & 0.0000 & -0.0370 & 0.0000 & -0.0145 \end{bmatrix}$$

$C=[0\ 1\ 0\ 0\ 0\ 0\ 0\ 0]$.

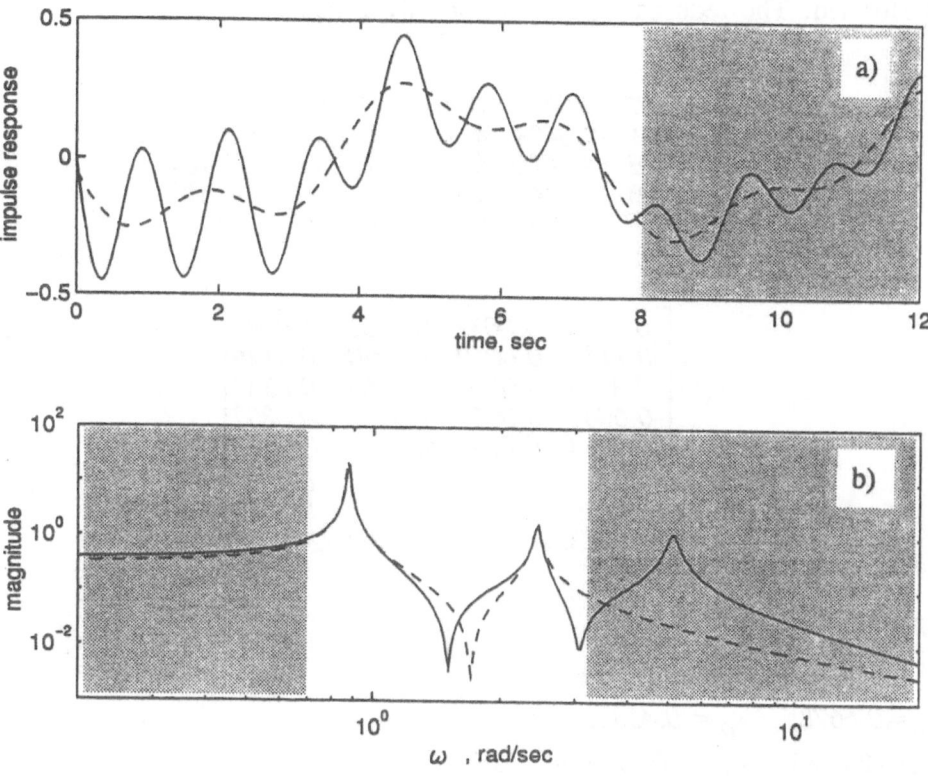

Figure 4.13: Impulse responses (a) and magnitudes of transfer function (b) of a simple system and its reduced models for $T=[0,\ 8]$ sec and $\Omega=[0.7,\ 3.2]$ rad/sec

The system poles are $\lambda_1=0.6687$, $\lambda_2=-1.7756$, $\lambda_{3,4}=-0.0150\pm j0.0886$, $\lambda_{5,6}=-0.3122\pm j4.4485$, $\lambda_{7,8}=-0.7257\pm j6.7018$, so there is one unstable pole. Colgren reduced the model by removing the unstable pole from the model and applying the reduction procedure in the infinite time interval, Ref.[16]. Then the unstable pole was added back to the reduced model. Here, the finite time balanced reduction is applied to the aircraft model which does not require the removal of the unstable pole. The time interval $T=[0, 4]$ sec is chosen, and model reduction from eight to five state variables within the time interval T is performed. The reduced model obtained (A_r, B_r, C_r) is

$$A_r = \begin{bmatrix} 0.6673 & -0.0218 & 0.0170 & -0.0213 & 0.0010 \\ -0.0049 & -0.0458 & -4.3161 & -0.2566 & -0.2227 \\ -0.0003 & 4.1620 & -0.5094 & -1.8500 & -0.6790 \\ -0.0203 & -0.3347 & 1.4769 & -1.5621 & -1.4259 \\ 0.0087 & -0.0007 & 0.7068 & -0.4785 & -1.0142 \end{bmatrix},$$

$$B_r = \begin{bmatrix} 0.0766 & 0.0032 & -0.0453 & 1.0041 \\ -0.1182 & -0.0019 & 0.0391 & 0.3129 \\ 0.8104 & -0.0001 & -0.2002 & 0.0382 \\ -0.6081 & -0.0069 & 0.1815 & 0.6852 \\ -0.4484 & 0.0042 & 0.1139 & -0.0650 \end{bmatrix}$$

$$C_r = [-1.0079 \quad -0.3324 \quad -0.8575 \quad -0.9270 \quad -0.4545]$$

with its poles

$$\lambda_{r1} = 0.6678, \quad \lambda_{r2} = -0.4533,$$
$$\lambda_{r3} = -2.0488, \quad \lambda_{r4,5} = -0.3148 \pm j4.5654.$$

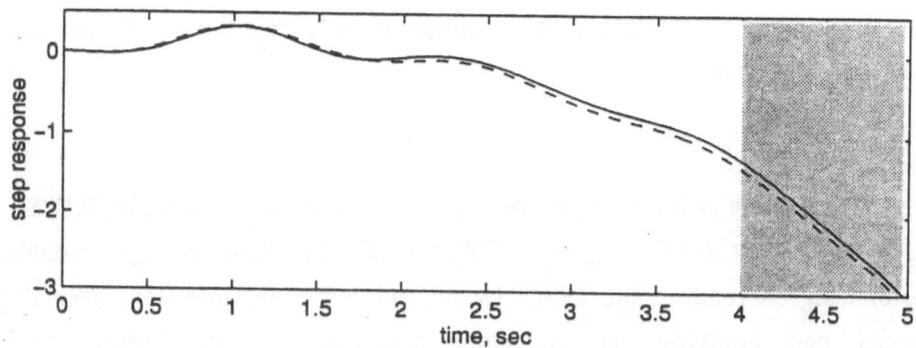

Figure 4.14: Step response to the first input of the Advanced Supersonic Transport and its reduced model for $t=[0, 1.4]$ sec

78

The reduction error is $\delta_b=0.0020$, and the optimality index is $\varepsilon_b=0.0085$, so that the reduced model obtained is near-optimal. Its step response within the time interval T due to the input 1 (dotted line) is close to the step response of the full-order model (solid line) in Fig.4.14, nonshaded area. The step responses of the reduced-order model due to input 2, 3, or 4, overlap the responses of the full-order model.

Truss. Model reduction of a flexible truss structure given in Fig.2.2 is investigated. The truss damping is proportional to the stiffness matrix, $d=0.0002k$, the vertical input force is located at node $n8$, and the vertical displacement of node $n4$ is the system output.

The reduction from 26 to 4 states is considered in three cases: Case 1, where the grammians and balanced representation are obtained in the time interval $T_1=[0, 2]$ sec, and frequency interval $\Omega_1=[0, 100]$ rad/sec; Case 2, the grammians are determined and balanced in the time interval $T_2=[0, 20]$ sec, and frequency interval $\Omega_2=[0, 3.5]$ rad/sec; Case 3, the time interval is $T_3=[0, 4]$ sec, and the frequency interval is $\Omega_3=[1, 100]$ rad/sec.

In Case 1 the frequency interval Ω_1 is large enough to include the system dynamics, thus it is considered a reduction in the time interval $T_1=[0, 2]$ sec. The impulse response and the magnitude of the transfer function of the full order system are shown in Fig.4.15a,b, solid line. Fig.4.15a shows that in the interval $T_1=[0, 2]$ sec (nonshaded) is dominated by the mode of frequency 12 rad/sec. The reduced model tries to minimize the error within this interval, thus it includes the mode of 12 rad/sec, as seen in Fig.4.15b.

In Case 2 the time interval is large enough to consider the reduction dependant on the frequency interval Ω_2 only. The reduction results are shown in Fig.4.16a,b. The transfer function Fig.4.16b shows that the reduced model fits the full model within the frequency interval $\Omega_2=[0,\ 3.5]$ rad/sec, indicated by the nonshaded part of the figure. Fig.4.16a show the impulse response of the full and the reduced model. It is seen that the reduced model does not include the important 12 rad/sec mode, as expected.

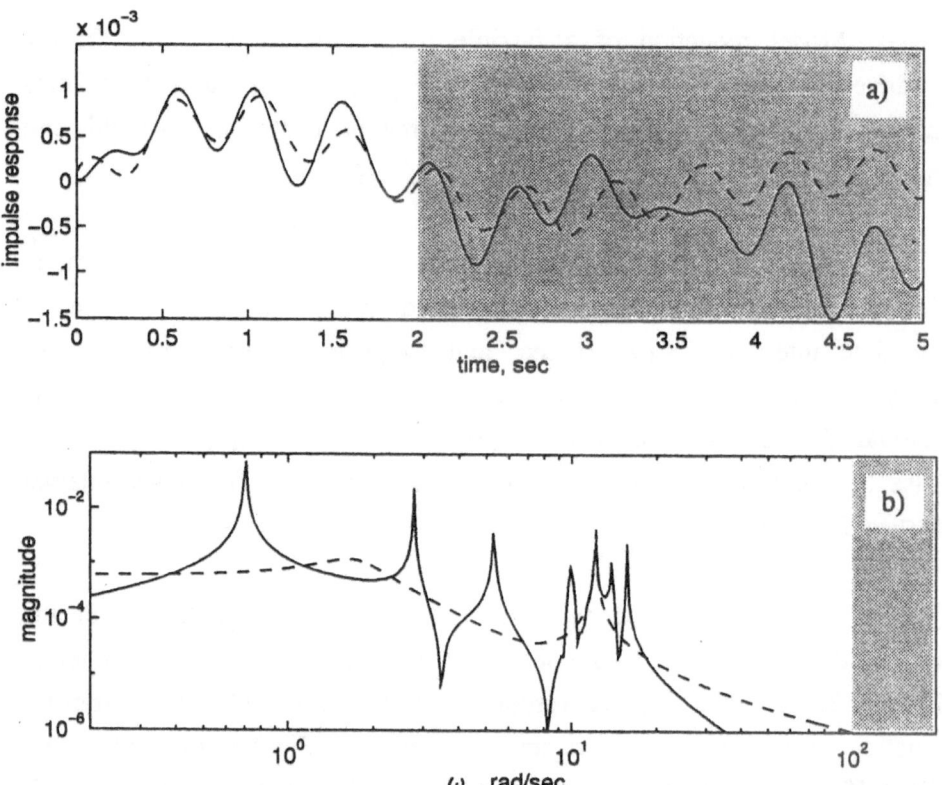

Figure 4.15: Impulse responses and magnitudes of transfer function of the truss structure and its reduced models for $T_1=[0,\ 2]$ sec and $\Omega_1=[0,\ 100]$ rad/sec

80

Case 3 performs reduction in the time interval $T_3=[0, 4]$ sec, and the frequency interval $\Omega_3=[1, 100]$ rad/sec. In this case the performance of the reduced model is expected to exclude the lowest mode, as illustrated in Fig.17a,b. The impulse response and the magnitude of the transfer function include the good fit for the modes of 2.8 rad/sec and 12 rad/sec in the interval Ω_3. These modes are the most visible within the interval T_3. The reduction and optimality indices are given in Table 4.4, showing that the reduction error was less than 2% in all cases.

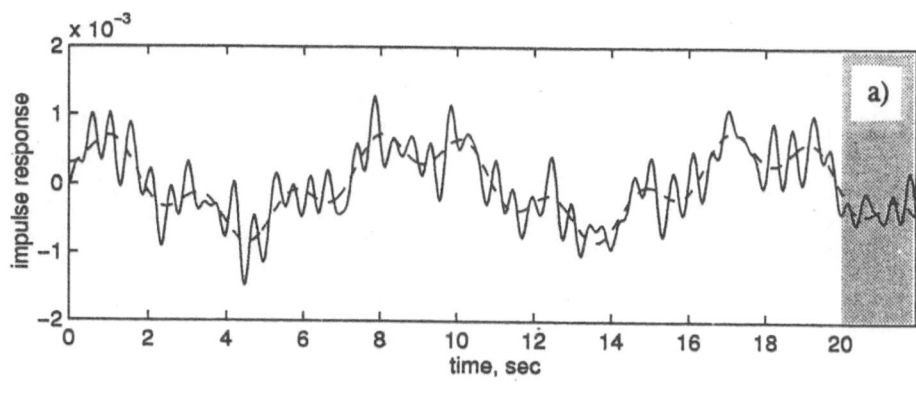

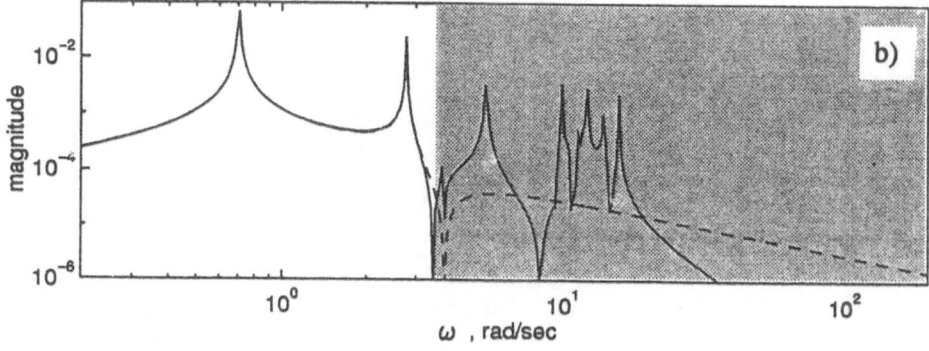

Figure 4.16: Impulse responses and magnitudes of transfer function of the truss and its reduced models for $T_2=[0, 100]$ sec, and $\Omega_2=[0, 3.5]$ rad/sec

Table 4.4. Reduction and optimality indices

	δ	ε
Case 1	0.0140	0.000008
Case 2	0.0002	0.000011
Case 3	0.0007	0.000432

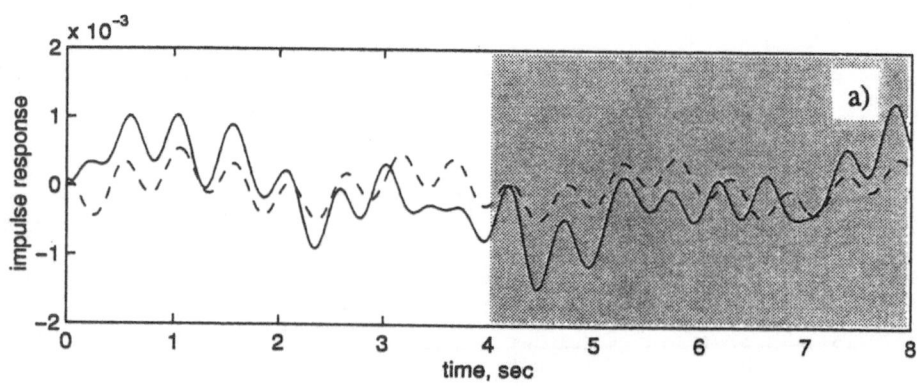

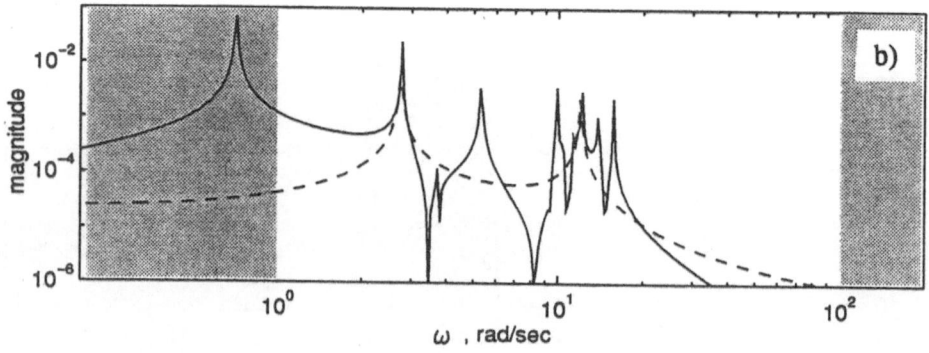

Figure 4.17: Impulse responses and magnitudes of transfer function of the truss structure and its reduced models for $T_1 = [0, 2]$ sec, and $\Omega_2 = [0, 3.5]$ rad/sec

4.6 Reduction of Systems with Integrators

For the purpose of reduction, a system with integrators is balanced, and its antigrammian is ordered increasingly:

$$\Pi=diag(\pi_1, \pi_2, \ldots, \pi_n) \tag{4.35}$$

where $\pi_i \geq 0$, $\pi_{i+1} \geq \pi_i$, $i=1,\ldots,n$, with the first m singular values at zero, $\pi_i=0$, $i=1,\ldots,m$. The system is reduced by truncating the last $n-k$ states of the balanced representation.

Example 4.13 A system with integrators and the state-space representation

$$A=\begin{bmatrix} -5 & 0 & 0 & 0 & 0 & 0 \\ 1 & 0 & 0 & 0 & 0 & 0 \\ 0.5 & 5 & -20 & -2 & 0 & 0 \\ 0 & 0 & 1 & 0 & 0 & 0 \\ 0 & 0 & 40 & 40 & -0.2 & -1 \\ 0 & 0 & 0 & 0 & 1 & 0 \end{bmatrix}$$

$B^T=[1\ 0\ 0\ 0\ 0\ 0]$, $C=[0\ 0\ 0\ 0\ 0\ 10]$, has one pole at zero. Although the system is controllable and observable, its controllability and observability grammians do not exist. The balanced representation is obtained from Eq.(3.50)

$$A_b=\begin{bmatrix} 0 & 0 & 0 & 0 & 0 & 0 \\ 0 & -0.1063 & -0.0549 & -0.0527 & 0.0091 & 0.0011 \\ 0 & -0.0549 & -0.1055 & -0.9949 & 0.0324 & 0.0040 \\ 0 & 0.0527 & 0.9949 & -0.8882 & 0.0328 & 0.0040 \\ 0 & 0.0091 & 0.0324 & -0.0328 & -5.1058 & -1.2517 \\ 0 & 0.0011 & 0.0040 & -0.0040 & -1.2517 & -1.9794 \end{bmatrix}$$

$$B_b^T=[20\ \ 13.9642\ \ 3.8892\ \ -3.2384\ \ -0.5972\ \ -0.0735],$$

$$C_b=[9.9964\ -13.9642\ -3.8892\ -3.2384\ \ 0.5972\ \ 0.0735]$$

83

and the balanced antigrammian $\pi=diag(0, 0.0011, 0.0139, 0.0169, 28.6333, 7330.82)$ satisfies Eq.(3.49). Its last two (largest) singular values indicate that the last two states can be truncated, and the system can be reduced to a system of order 4. Indeed, as shown in Fig.4.18, the output of the reduced model, obtained by truncation is close to the output of the full-order model, as is illustrated by the transfer functions of the full (solid line) and the reduced order model (dashed line).

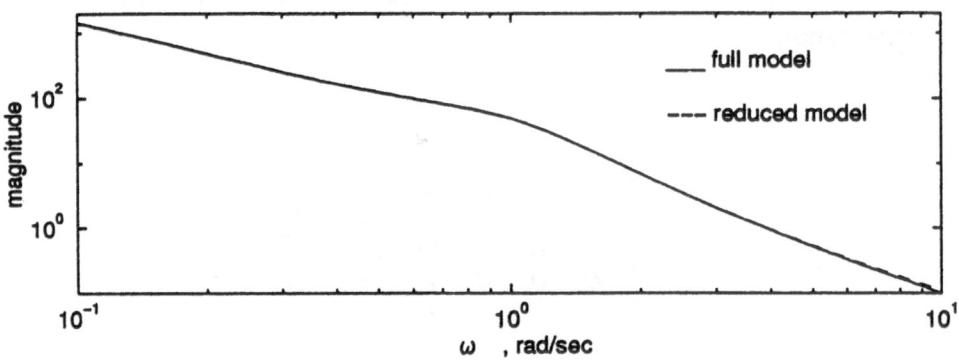

Figure 4.18: Magnitudes of transfer function of the full-order and reduced system with integrators

Example 4.14 Consider the Deep Space Network antenna model which has three poles at zero. The state-space representation of the rate loop model has $n=67$ states, including three states with poles at zero. This model is reduced using antigrammians: their singular values are shown in Fig.4.19. By deleting the states with the largest singular values one obtains the 30-state model. The reduced model preserves properties of the full-model, as it is shown with the magnitude of the transfer function in Fig.4.20.

4.7 Reduction Using Transfer Function

Two indices used for the model reduction, Hankel singular values and component costs, were obtained in time domain from the state-space representation (A,B,C). However, the state-space representation of a system is not always available. In many cases, a transfer function, either from measurements or for analysis, can be used. The determination of indices in frequency domain from the transfer function, is presented in this section.

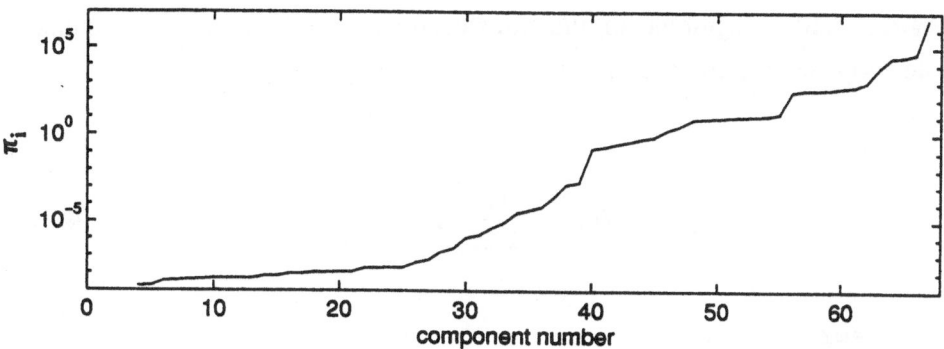

Figure 4.19: Singular values of the antigrammians for the rate loop model

The transfer function $H=C(j\omega I-A)^{-1}B$ for the flexible structure in modal coordinates of form 1 or 2 has the following form

$$H(\omega)= \sum_{i=1}^{n_2} H_i(\omega), \quad H_i(\omega)=C_i(j\omega I-A_i)^{-1}B_i \tag{4.36}$$

For $\omega=\omega_i$ one obtains $\|H_k(\omega_i)\| \ll \|H_i(\omega_i)\|$ for $k\neq i$, therefore $H(\omega_i)\cong H_i(\omega_i)$. From the modal form 1, after some algebra, one finds that

85

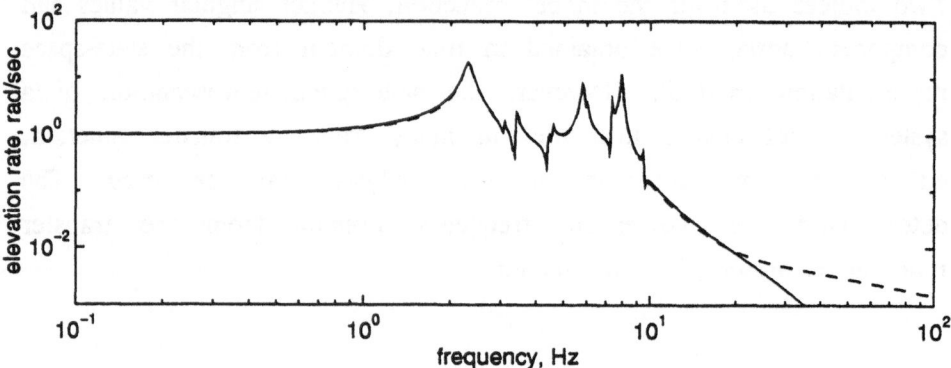

Figure 4.20: Magnitude of the transfer function of the full and reduced rate loop model, for the elevation rate output

$$H_i(\omega_i) = \frac{c_i b_i}{2\zeta_i \omega_i^2} \; , \tag{4.37}$$

and $c_i = \omega_i c_{ri} - j c_{di}$. Define

$$cos\psi_i = \frac{\|c_i b_i\|}{\|c_i\| \|b_i\|}, \tag{4.38}$$

which is a measure of collinearity of c_i and b_i. The norm of $H_i(\omega_i)$, denoted by h_i, is the norm of the output at frequency $\omega = \omega_i$

$$h_i = \|H_i(\omega_i)\| = \frac{\|b_i\| \|c_i\|}{2\zeta_i \omega_i^2} \; cos\psi_i \tag{4.39}$$

For the orthogonal c_i and b_i one obtains $h_i = 0$, whereas for a single-input and/or a single-output, or collocated sensors and actuators, $cos\psi_i = 1$.

As mentioned before, flexible structures in modal coordinates have controllability and observability grammians diagonally dominant, that is $W_c \cong diag(W_{ci})$ and $W_o \cong diag(W_{oi})$ where

$$W_{ci} = I_2 \frac{\|b_i\|^2}{4\zeta_i\omega_i}, \qquad W_{oi} = I_2 \frac{h_i^2\zeta_i\omega_i}{\|b_i\|^2} \qquad (4.40)$$

and the balanced grammians are computed simply as $W_{cb} = W_{ob} = (W_cW_o)^{1/2} = diag(\gamma_i^2 I_2)$. From this fact, and from Eq.(4.40) it follows that for flexible structures the square of the ith Hankel singular value equal to one-half the norm of the output at the ith resonant frequency,

$$\gamma_i^2 = \frac{h_i}{2} = \frac{\|b_i\|\|c_i\|\cos\psi_i}{4\zeta_i\omega_i^2} = \frac{\sqrt{b_ib_i^T(c_{di}^Tc_{di} + \omega_i^2 c_{ri}^Tc_{ri})}}{4\zeta_i\omega_i^2}\cos\psi_i \qquad (4.41)$$

Figure 4.21 shows the determination of the Hankel singular value of the first mode, γ_1^2, from the magnitude of the transfer function for a single-input and single-output system, $\gamma_1^2 = 0.5h_1 = 0.5\|H(\omega_1)\|$.

The cost of the ith modal coordinate determined from Eq.(4.12) is as follows

$$\sigma_i^2 = -tr(A_i\gamma_i^4) = 0.5\zeta_i\omega_ih_i^2 \qquad (4.42)$$

Denoting the half power frequency $\Delta\omega_i = 2\zeta_i\omega_i$, see Refs.[15], and [27], one may rewrite (4.42) as

$$\sigma_i = \frac{h_i\sqrt{\Delta\omega_i}}{2} = \frac{\sqrt{b_ib_i^T(c_{di}^Tc_{di} + \omega_i^2 c_{ri}^Tc_{ri})\Delta\omega_i}}{4\zeta_i\omega_i^2}\cos\psi_i \qquad (4.43)$$

Thus, unlike Hankel singular values, the cost consists of a product of the resonance amplitude and the resonance width.

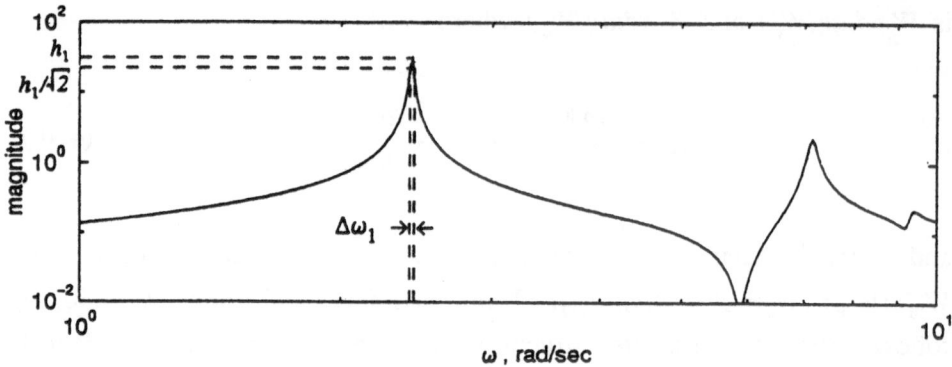

Figure 4.21: The determination of the Hankel singular value and component cost from a transfer function

The half power frequency can be determined from the magnitude of transfer function, see Fig.4.21. For each resonance peak the resonance frequency (ω_i) and maximum amplitude (h_i) are determined. The half power frequency $\Delta\omega_i$ is the frequency bandwidth of the transfer function at the level of $h_i/\sqrt{2}$.

The results (4.41) and (4.43) show that for small damping and separated natural frequencies, the reduction technique in modal or balanced coordinates involves truncating the variables with the smallest resonance peaks (γ truncation), or with the smallest product of the resonance peaks and resonance width (σ truncation). These techniques were commonly used before the Hankel singular values or component costs were introduced. However, both balanced and modal techniques develop a theoretical background and new analytical tools for the already-in-use techniques.

Example 4.15 A simple system from Fig.2.1 is investigated. The system parameters are as follows: $m_1=m_2=m_3=2$, $k_1=20$, $k_2=k_4=50$, $k_3=100$, $d_i=0.001k_i$, $i=1,2,3,4$, with single input applied to the first mass,

$u=f_1$, and two outputs, displacements of the second mass $(y_1=q_2)$ and velocity of the first mass $(y_2=v_1)$. Table 4.5 gives the system Hankel singular values, γ_i, obtained analytically, as well as the Hankel singular values, γ_{ei}, obtained from the frequency response function. It shows that the transfer function gives a good approximation of Hankel singular values. The costs, σ_i, obtained analytically and the cost, σ_{ei}, from the transfer function are also given in Table 4.5, showing that $\sigma_i \cong \sigma_{ei}$ for $i=1,...,6$.

Next, the flexible truss in Fig.2.2 with 26 state variables, and with a single input and single output (both collocated at node $n8$, vertical direction) is considered. The sensor and actuator are placed such that all Hankel singular values are equal: $\gamma_i=1$, for $i=1,...,26$, i.e., all states are equally controlled and observed. The plots of Hankel singular values are given in Fig.4.22. The analytical Hankel singular values (solid line) and those obtained from the transfer function (dashed line) are nearly the same.

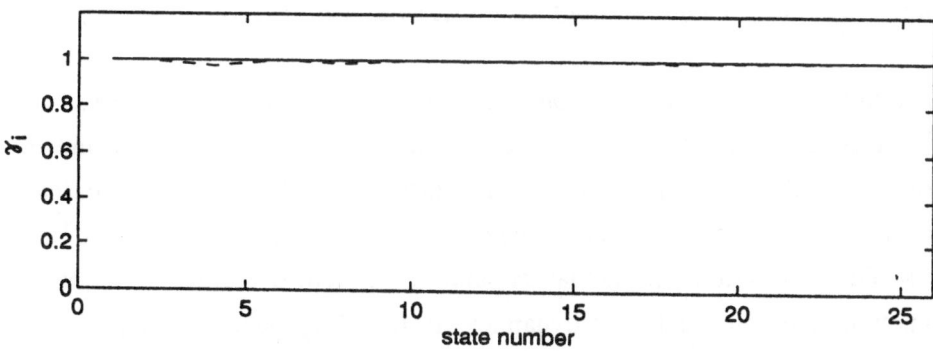

Figure 4.22: A comparison of the Hankel singular values: analytical and from the transfer function

Table 4.5. Actual and estimated Hankel singular values and costs

i	γ_i	γ_{ei}	σ_i	σ_{ei}
1	3.1230	3.1228	12.4475	12.4487
2	3.1226	3.1228	12.4493	12.4487
3	1.7591	1.7592	5.7215	5.7072
4	1.7591	1.7592	5.6922	5.7072
5	0.2745	0.2842	0.1800	0.1928
6	0.2744	0.2842	0.1796	0.1928

4.8 Reduction with Actuator and Sensor Dynamics

A flexible structure in a closed-loop configuration is typically equipped with sensors and actuators. Does the presence of sensors and actuators impact the dynamics of a flexible structure, and consequently the process of model reduction? This question will be answered for three important cases: sensors and actuators in cascade connection with a structure, accelerometer sensors, and for the proof-mass actuators attached to a structure.

4.8.1 Actuators and Sensors in a Cascade Connection

Without loosing generality, only actuator dynamics is considered, and the reconstruction of the flexible structure reduction indices from the joint actuator-structure indices is discussed. The sensor cascade connection is similar to the actuator case. The system considered is a cascade connection of actuators and a structure. Thus its transfer function, including actuator dynamics, is $H=H_sH_a$, where H_s and H_a are transfer functions of the structure and actuators, respectively. Define

$$cos\phi = \frac{\|H\|}{\|H_s\|\|H_a\|} \tag{4.44}$$

where the Froebenius norm is used. The value of $cos\phi$ is a measure of the collinearity of H_a and H_s. If the actuator transfer function is collinear with the structure transfer function, then $cos\phi=1$, whereas orthogonal H_a and H_s result in $cos\phi=0$, and $H=0$. From Eq.(4.44) one obtains

$$\|H_s\| = \frac{\|H\|}{\|H_a\|cos\phi} \tag{4.45}$$

Since $\|H_s(\omega_i)\| = h_{si} = 2\gamma_{si}^2$, and $\|H(\omega_i)\| = h_i = 2\gamma_i^2$, see Eq.(4.41), consequently from Eq.(4.45) it follows that

$$\gamma_{si}^2 = \frac{\gamma_i^2}{\alpha_i}, \quad \alpha_i = \|H_{ai}\|cos\phi_i, \quad i=1,...,n \tag{4.46}$$

with notation $H_{ai}=H_a(\omega_i)$, $cos\phi_i=cos\phi(\omega_i)$, and ω_i being the ith resonance frequency. The Hankel singular values of the structure are obtained from the Hankel singular values of the structure with actuators by scaling the latter by the amplitude of actuator outputs at the i-th resonance frequency, and collinearity measurement of H_a and H_s.

Assuming that $\|H_a(\omega)\|$ has no abrupt changes in the neighborhood of the natural frequencies, we have in this case $\Delta_i \cong \Delta_{si}$. Therefore the component cost is similarly scaled as the Hankel singular values

$$\sigma_{si} = \frac{\sigma_i}{\alpha_i}, \quad i=1,...,n \tag{4.47}$$

Eq.(4.46) and Eq.(4.47) show that every Hankel singular values and component costs for the structure alone are obtained by scaling related Hankel singular values and component costs of the combined structure-actuator system. If $cos\phi_i$ is constant for $i=1,...,n$, then the squares of Hankel singular values and the component costs are scaled by the inverse of the actuator gain.

The determination of $cos\phi$ from Eq.(4.44) using test data may be difficult, if not impossible, since H_s is not known explicitly during testing. The difficulty can be avoided if we notice that introducing (4.37) into (4.44) yields

$$cos\phi_i = \frac{\|c_{si}b_{si}H_{ai}\|}{\|c_{si}b_{si}\|\,\|H_{ai}\|} \qquad (4.48)$$

Here, only the actuator transfer function and actuator-sensor location are involved. For a single-input system $cos\phi_i = \|c_{si}H_{ai}\|/(\|c_{si}\|\,\|H_{ai}\|)$, and for a single-input and single-output system $cos\phi = 1$.

Example 4.16 The system from Fig.2.1, with an actuator attached to the single input, is examined. The system parameters are as follows: $m_1 = 11$, $m_2 = 5$, $m_3 = 10$, $k_1 = 10$, $k_2 = 50$, $k_3 = 55$, $k_4 = 10$, $d_i = 0.001k_i$, $i = 1,2,3,4$, with single input applied to the first mass, and a single output, the displacement of the third mass. The actuator transfer function is $H_a = 1/(1+s)$. The transfer function of the combined structure and actuator is plotted in Fig.4.23 (solid line), including the structure transfer function (dotted line), and the actuator transfer function (dashed line). The Hankel singular values γ_{si} and the costs σ_{si} of the structure are given in Table 4.6. For this system $cos\phi_i = 1$ for $i = 1,2,3$, and $\alpha_i = 0.0217$, $\alpha_2 = 0.00972$, and $\alpha_3 = 0.00094$. The reconstructed structure Hankel singular values and costs, presented in Table 4.6, are determined from the combined structure and actuator transfer function using Eqs.(4.46) and (4.47). One can see that $\gamma_{si} \cong \gamma_{sri}$ and $\sigma_{si} \cong \sigma_{sri}$ for $i = 1,...,6$.

Table 4.6. Actual and reconstructed Hankel singular values

i	γ_{si}	γ_{sri}	σ_{si}	σ_{sri}
1	14.2084	14.2115	0.3076	0.3077
2	14.1836	14.1928	0.3079	0.3081
3	0.8228	0.8221	0.00799	0.00798
4	0.8188	0.8207	0.00801	0.00804
5	0.0180	0.0180	0.00002	0.00002
6	0.0179	0.0178	0.00002	0.00002

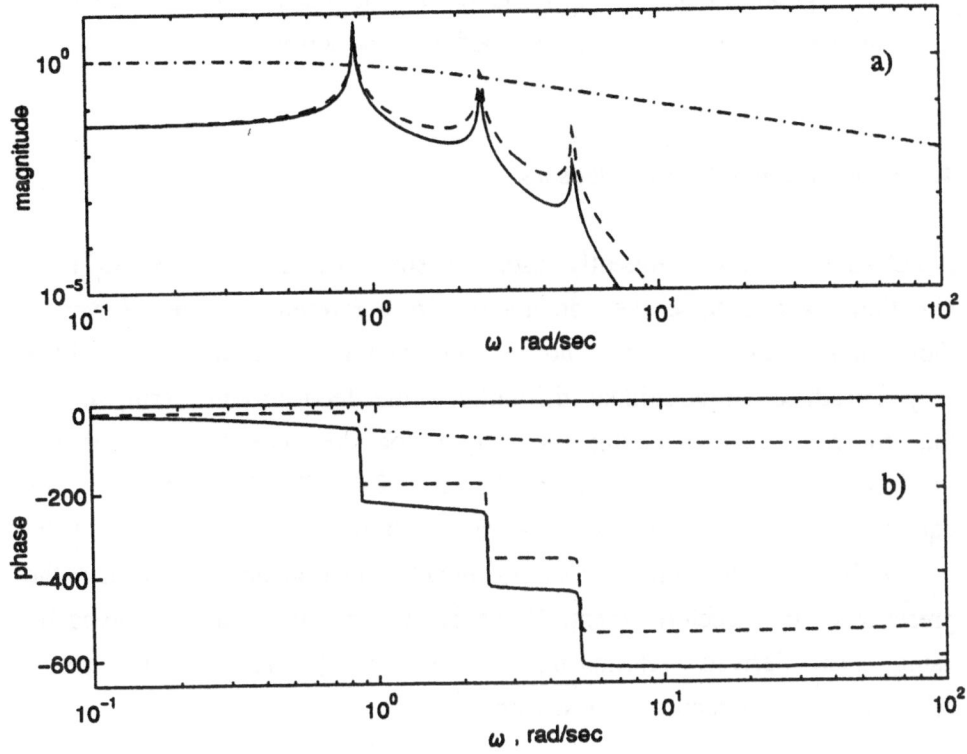

Figure 4.23: Transfer function of the simple system (dotted line), the actuator (dashed line), and their combination (solid line): (a) magnitude, (b) phase

Example 4.17 The flexible truss in Fig.2.2 with 26 state variables, and with a single input and single output (such that the Hankel singular values of all states are equal to 1) is considered. An actuator with the transfer function $H_a = 1/(s+1)$ is applied. The transfer function of the truss and actuator connected in series is plotted in Fig.4.24 (solid line), including the truss transfer function in dotted line and and the actuator transfer function (dashed line). The Hankel singular values of the truss are all equal to 1 (see solid line in Fig.4.25), while the Hankel singular values of the truss and actuator combination are plotted in the dotted line. The reconstructed truss Hankel singular values are represented in the dashed line (it overlaps most of the solid line), indicating good reconstruction results.

4.8.2 Accelerometers as Sensors

Accelerometers are frequently used as structural sensors due to their simplicity, and because they do not require a reference frame. However, they amplify high frequency noise, and from the standpoint of analysis they introduce an additional difficulty - the feed-through term, D, in the output equation, giving the output for the accelerometer sensors $y = Cx + Du$, rather than $y = Cx$, as in Eq.(2.2a). The difficulty with this equation follows from the fact that the grammians do not depend on the matrix D. Thus the grammian-based model reduction does not reflect the presence of the accelerometers. However, this problem can be solved by using just derived relationships (4.46) or (4.47) for the series connection of a structure and sensors.

A structure with the accelerometers can be considered as a structure with rate sensors in a cascade connection with differenciators placed after each sensor (the derivative of a rate gives an acceleration). Thus, by using the above equations, the Hankel singular values of a structure equipped with accelerometers can be determined from the Hankel singular values of a structure with rate sensors. For simplicity of notation consider a structure with a single accelerometer. Denote (A_r, B_r, C_r), and $H_r = C_r(sI - A_r)^{-1}B_r$ the state space triple and the transfer function, respectively, of the structure with a rate sensor. The transfer function H_a of the structure with an accelerator is therefore

$$H_a = j\omega H_r \qquad (4.49)$$

94

According to Eq.(4.44) one obtains $cos\phi=1$, and since $\|j\omega_i\|=\omega_i$, thus from (4.46) it follows that

$$\gamma_{ia}^2 \cong \omega_i \gamma_{ir}^2, \qquad (4.50)$$

$i=1,...,n$, where ω_i is the i-th natural frequency. Similarly, from Eq.(4.47) one obtains for $i=1,...,n$

$$\sigma_{ia} \cong \omega_i \sigma_{ir}, \qquad (4.51)$$

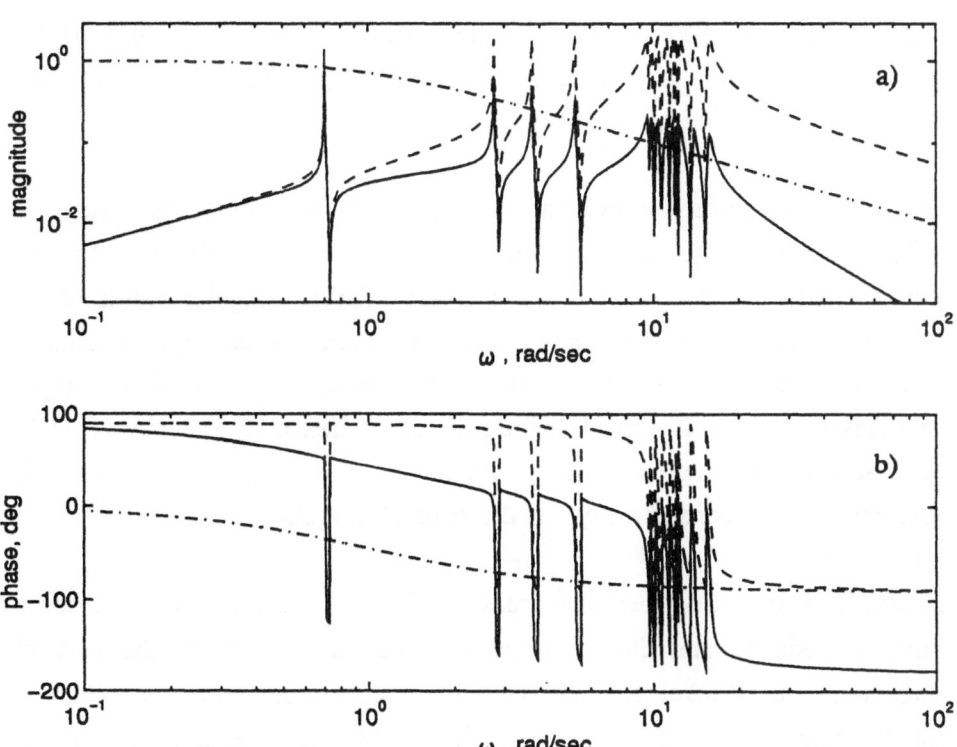

Figure 4.24: Transfer function of the truss structure (dotted line), the actuator (dashed line), and their combination (solid line): (a) magnitude, (b) phase

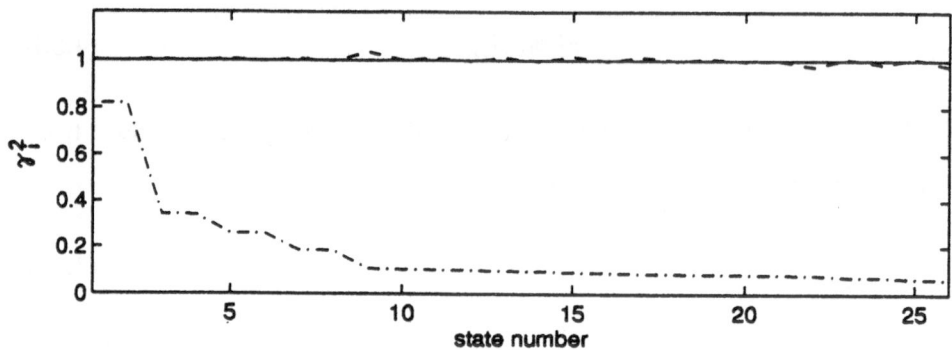

Figure 4.25: Hankel singular values of the truss structure (solid line), the truss and actuator combined (dotted line), and the reconstructed Hankel singular values (dashed line)

The above equations show that the square of the ith Hankel singular value of a structure with an accelerometer sensor is obtained by the multiplication of the square of the ith Hankel singular value of a structure with a rate sensor by the ith natural frequency. A similar conclusion applies to the costs. The importance of this simple relationship is that it solved the unwanted presence of the matrix D in the state-space representation, which is not represented in the grammians, and consequently, in the Hankel singular values.

Example 4.18   Consider the truss in Fig.2.2 with the vertical force input at node $n7$ and the acceleration output at node $n8$ in the vertical direction. The Hankel singular values for this system can be obtained by combining in series the truss with the rate measurement and a differenciator (with small time constant). As a differenciator we used a system with the transfer function $s/(0.001s+1)$. The Hankel singular values for this system are exactly equal to the truss with the acceleration output, and are shown in Fig.4.26, solid line. These Hankel singular values were also obtained from the approximate equation

96

(4.50), and are shown in the same figure, dashed line. The approximate values are sufficiently close to the exact ones.

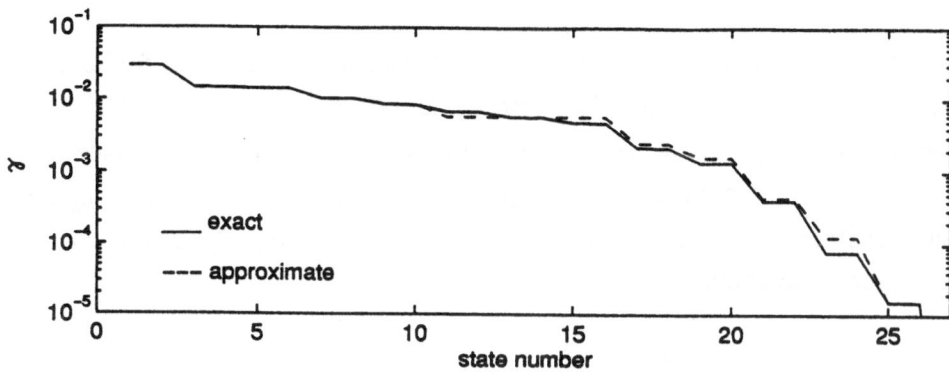

Figure 4.26: Hankel singular values for the truss with the accelerometer sensor.

4.8.3 Proof-Mass Actuators

Proof-mass actuators (PMAs) are widely used in structural dynamics testing. In most investigations, however, the actuator dynamics are not included in the model. Here, the relationship between the Hankel singular values of a structure with a proof-mass actuator and the Hankel singular values of the structure alone (i.e., with an ideal actuator) is studied, and the influence of the PMA on model reduction is analyzed and demonstrated.

Two types of PMAs are typically used. Both consist of mass m, spring with stiffness k, and they are attached to a structure at node, say, na. The first one, type A PMA, is a reaction-type force actuator, see Ref.[123]. It generates a force by reacting against the mass m, thus force f acts on the structure, and $-f$ on the mass m (Fig.4.27a). The second one, type B PMA, is a centrifugal actuator, where force is

97

proportional to the square of the excitation frequency. The force acts on mass m exclusively (Fig.4.27b). It is assumed that the stiffness of the PMA is much smaller than the dynamic stiffness of the structure (often it is zero).

Structure with Type A PMA   Let us consider a structure without an actuator, shown in Fig.4.27c. Its dynamics is described by Eq.(2.1), where u is the force acting on the structure, q is its displacement vector, y is its output vector, M_s, D_s, and K_s, are the mass, damping, and stiffness matrices of the structure, respectively, B_s is the matrix of actuator location,

$$B_s = [0 \ 0 \ .. \ 0 \ 1 \ 0 \ .. \ 0]^T \qquad (4.52)$$

with a nonzero term at location na, and $C_s = C_{oq}$ is a matrix of sensor location. Denote $G_s(\omega) = -\omega^2 M_s + j\omega D_s + K_s$, $j = \sqrt{-1}$, then the structure transfer function (the dynamic elasticity of the structure) is

$$H_s = C_s G_s^{-1} B_s \qquad (4.53)$$

The dynamic stiffness of a structure at the actuator location is defined as

$$k_s = \frac{1}{B_s^T G_s^{-1} B_s} \qquad (4.54)$$

The dynamic stiffness is the inverse of the frequency response function, i.e., displacement to force relationship, at the actuator location. At zero frequency, it reduces to the stiffness constant at the actuator location. For a structure with type A PMA denote q_a, m, k, d, the displacement, mass, stiffness, and damping of the actuator, and define the displacement vector $q_c^T = [q_s^T, q_a]$. The combined structure and actuator equation is

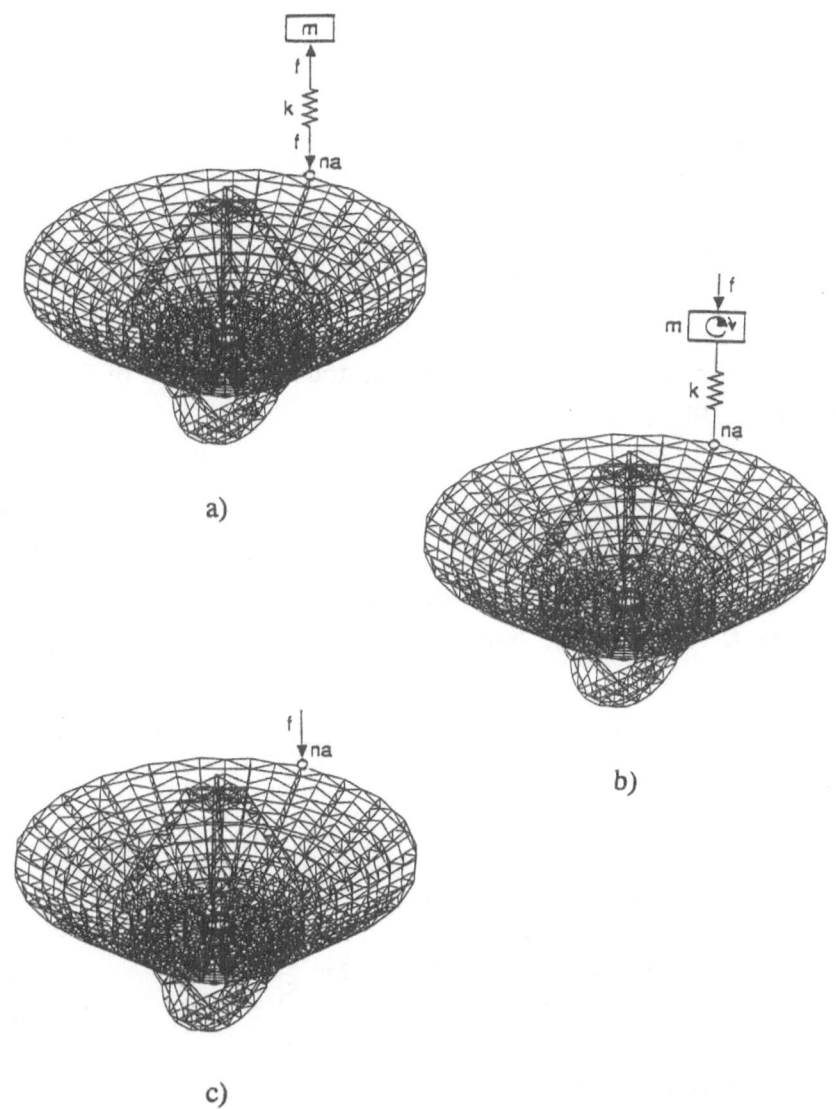

Figure 4.27: A structure with: (a) type A actuator, (b) type B actuator, (c) ideal actuator

$$M_c \ddot{q}_c + D_c \dot{q}_c + K_c q_c = B_c f, \qquad y_c = C_c q_c \qquad (4.55)$$

where

$$M_c = \begin{bmatrix} M_s & 0 \\ 0 & m \end{bmatrix}, \qquad D_c = \begin{bmatrix} D_s & 0 \\ 0 & d \end{bmatrix}, \qquad B_c = \begin{bmatrix} B_s \\ -1 \end{bmatrix}, \qquad C_c = [C_s \ 0], \qquad (4.56a)$$

and

$$K_c = \begin{bmatrix} K_s + kB_s B_s^T & -kB_s \\ -kB_s^T & k \end{bmatrix}, \qquad D_c = \begin{bmatrix} D_s + dB_s B_s^T & -dB_s \\ -dB_s^T & d \end{bmatrix} \qquad (4.56b)$$

Denote $G_c(\omega) = -\omega^2 M_c + j\omega D_c + K_c$, then the transfer function of the structure with type A PMA is

$$H_c = C_c G_c^{-1} B_c \qquad (4.57)$$

and the relationship between the transfer function of the structure with type A PMA and the transfer function of the structure alone is

$$H_c = \alpha_c H_s, \qquad \alpha_c = \frac{1}{1 + \beta - \rho^2} \qquad (4.58a)$$

where

$$\rho = \frac{\omega_0}{\omega}, \qquad \omega_0 = \sqrt{\frac{k}{m}}, \qquad \beta = \frac{k}{k_s}. \qquad (4.58b)$$

Eq.(4.58) is obtained as follows. Denote

$$H_o = (G_s + kB_s B_s^T)^{-1}, \quad g = k - \omega^2 m, \quad \text{and} \quad \Delta = g - k^2 B_s^T H_o B_s \qquad (4.59)$$

then

$$G_c = \begin{bmatrix} K_s - \omega^2 M_s + j\omega D_s & -kB_s \\ -kB_s^T & g \end{bmatrix} = \begin{bmatrix} -H_o^{-1} & -kB_s \\ -kB_s^T & \Delta + -k^2 B_s^T H_o B_s \end{bmatrix} \qquad (4.60)$$

By applying the partitioned inverse formula given in Ref.[67] to Eq.(4.59) one obtains

$$G_c^{-1} = \Delta^{-1} \begin{bmatrix} \Delta H_o + k^2 H_o B_s B_s^T H_o & -kH_o B_s \\ -kB_s^T H_o^T & 1 \end{bmatrix} \qquad (4.61)$$

Next, from eqs. (4.56b) and (4.61) one obtains

$$H_c = C_s((H_o + k^2 \Delta^{-1} H_o B_s B_s^T H_o) B_s - \Delta^{-1} k H_o B_s)$$

$$= C(I - k\Delta^{-1}(I - kH_o B_s B_s^T)) H_o B_s \qquad (4.62)$$

and from (4.59)

$$H_o = (I - \frac{kG^{-1} B_s B_s^T}{1 + kB_s^T G^{-1} B_s}) G^{-1} \qquad (4.63)$$

thus

$$H_o B_s = \frac{G_c^{-1} B_s}{1 + \beta} \qquad (4.64)$$

Combining Eqs.(4.57), (4.61) and (4.64) one obtains Eq.(4.58).

For the case where the conditions

$$\omega \gg \omega_o \quad (\rho \ll 1) \quad \text{and} \quad k \ll k_s \quad (\beta \ll 1) \qquad (4.65)$$

are satisfied, one obtains $\alpha_c \cong 1$. In this case the transfer function of the system with the PMA is approximately equal to the transfer function of the system without the PMA. For example, for a PMA with a soft centering spring with low damping the above conditions are satisfied.

Consider Hankel singular values of a structure alone (γ_{si}) and of a structure with a type A PMA (γ_{ci}). They are related as follows

$$\gamma_{si} = \frac{\gamma_{ci}}{\alpha_{ci}}, \qquad i=1,\dots,n \qquad (4.66)$$

where

$$\alpha_{ci} = \alpha_c(\omega_i) = \frac{1}{1+\beta_i \cdot \rho_i^2}, \qquad (4.67a)$$

where $\rho_i = \rho(\omega_i) = \omega_o/\omega_i$, and

$$\beta_i = \beta(\omega_i) = \frac{k}{k_{si}} \qquad \text{and} \qquad k_{si} = k_s(\omega_i) = \frac{1}{B_s^T G_s^{-1}(\omega_i) B_s} \qquad (4.67b)$$

The variable k_{si} is the ith modal stiffness of the structure.

The relationship (4.66) can be proved as follows. For small damping and distinct natural frequencies the Hankel singular values of a flexible structure are given by Eq.(4.41). Introducing Eq.(4.58) to Eq.(4.41) one obtains Eq.(4.66).

In addition to conditions (4.65), consider the following ones

$$\omega_o \ll \omega_1 \quad \text{and} \quad k \ll \min_i k_{si} \qquad (4.68)$$

where ω_1 is the fundamental (lowest) frequency of the structure. These conditions say that the actuator natural frequency should be significantly lower than the fundamental frequency of the structure, and that the actuator stiffness should be much smaller than the dynamic stiffness of the structure at any frequency of interest. If the aforementioned conditions are satisfied, one obtains $\alpha_{ci} \cong 1$ for $i=1,...,n$, thus the Hankel singular values of the structure with the PMA are equal to the Hankel singular values of the structure without the PMA. The controllability and observability properties of the system are preserved. In particular, the presence of the PMA will not affect the model order reduction. Note also that for many cases, whenever the first condition of Eq. (4.65) is satisfied, the second condition Eq.(4.68) is satisfied too.

Example 4.19 A truss structure from Fig.2.2 is investigated. A vertical force is applied at node $n8$, and the output is measured at node $n4$ in the vertical direction. The magnitude of the transfer function of this structure is presented in Fig.4.28, solid line. The PMA is applied at node $n8$, with mass $m=200$ lbs sec²/in, and stiffness $k=0.6$ lbs/in, hence its natural frequency is $\omega_o=0.0548$ rad/sec. The actuator amplification factor α_c is plotted in Fig.4.29. The figure indicates that for $\omega \gg \omega_o$ the amplification is equal to 1. The plot of the magnitude of the transfer function for the structure with the PMA is shown in Fig.4.28 as a dashed line. The figure shows perfect overlapping of the transfer functions for $\omega \gg \omega_o$. Hankel singular values of the structure without and with the PMA are plotted in

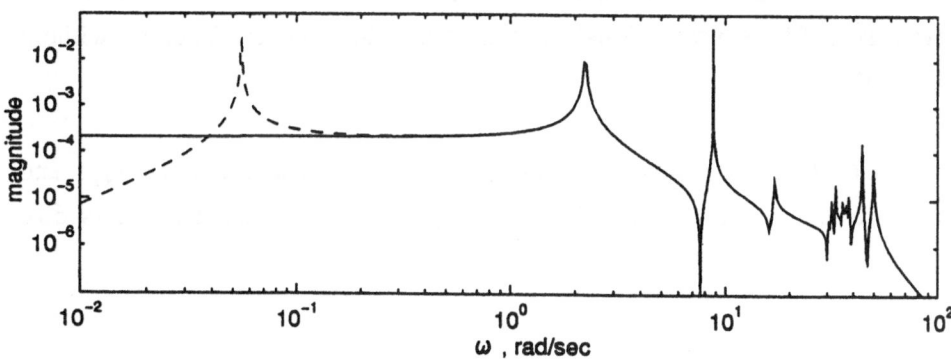

Figure 4.28: Magnitude of transfer function of a truss with an ideal actuator (solid line), and with type A actuator (dashed line)

Fig.4.30, solid and dashed line, respectively. Observe that the Hankel singular values are the same in both cases, except for the first one, which is related to the PMA itself.

Structure with Type B PMA This configuration is shown in Fig.4.27b. The force acting on mass m is proportional to the squared frequency

$$f = \kappa \omega^2 \tag{4.69}$$

where κ is a constant. The relationship between transfer functions of a structure without and without a type B PMA is as follows

$$H_c = \alpha_c H_s, \quad \alpha_c = \frac{\kappa \omega_0^2}{1 + \beta - \rho^2} \tag{4.70}$$

which can be proved by defining the displacement vector $q_c^T = [q_s^T, q_a]$, with matrices M_c, K_c, and C_c as in Eq.(4.56a), and $B_c^T = [0 \quad \kappa \omega^2]$. In this case one obtains G_c^{-1}, as in Eq.(4.61), and consequently (4.70). For the conditions (4.68) satisfied, one obtains the constant α_c

$$\alpha_c = -\kappa \omega_0^2 \tag{4.71}$$

The above result shows that the structural transfer function with the type B PMA is proportional to the structural transfer function without the PMA.

Consider the Hankel singular values of a structure alone (γ_{si}) and of a structure with a type B PMA (γ_{ci}). They are related as in (4.66), with

$$\alpha_{ci} = \frac{\kappa \omega_0^2}{1 + \beta_i - \rho_i^2} \tag{4.72}$$

With the conditions in Eq.(4.68) satisfied one obtains $\alpha_{ci} = \kappa \omega_0^2$ for

104

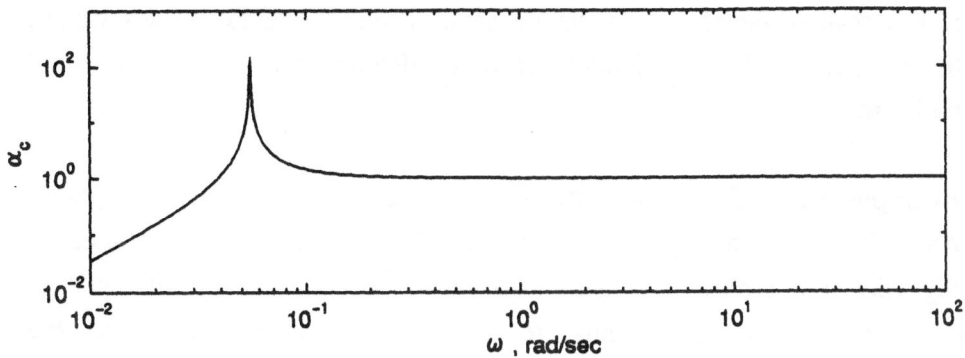

Figure 4.29: Amplification factor of type A actuator

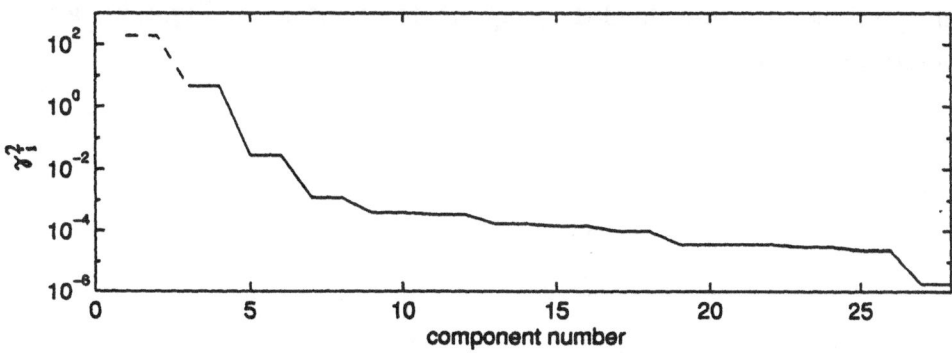

Figure 4.30: Hankel singular values of the truss with an ideal actuator (solid line), and with type A actuator (dashed line)

$i=1,...,n$, thus the Hankel singular values of the structure with the type B PMA are proportional to the Hankel singular values of the structure without the PMA. This scaling does not influence the results of reduction or model order determination procedures, since the procedures are based on ratios of Hankel singular values rather than their actual values.

In applications where measurement noise is unavoidable, care should be taken in the selection of the scaling factor. For instance, if the scaling factor is too small, small Hankel singular values cannot be detected, and the reduction or order determination procedure could be biased.

Example 4.20   The system described in Example 4.18 is investigated. Amplification factor α_c defined in Eq.(4.70) is the same as plotted in Fig.4.29, but scaled with the factor *0.003*. It is constant and equal to $\omega_o^2 = 0.003$ for $\omega \gg \omega_o$. Thus, to obtain the transfer function of the structure, the transfer function of the structure and the PMA are divided by ω_o^2. The same is valid for the Hankel singular values. Plots of magnitudes of the transfer functions of the structure (solid line) and the structure with the PMA divided by ω_o^2 (dashed line) are exactly the same as in Fig.4.28. Both plots overlap for $\omega \gg \omega_o$. Similarly, plots of Hankel singular values of the structure and Hankel singular values of the structure divided by ω_o^2 are the same as in Fig.4.30, and overlap for $\omega \gg \omega_o$.

Chapter 5

Balanced Sensor and Actuator Placement

For the purposes of flexible structure testing and control it is useful to investigate possible sensor and/or actuator locations, and to evaluate their impact on the dynamic test results, or the closed-loop performance. Of course, not always a freedom of choice of the locations is given, and even so, the choice may be limited. But, given the choice, two problems will surface. In the first one, the grammian assignment problem, the sensor/actuator locations of an open loop-system are determined to meet (assign) the specified controllability and observability requirements. In the second one, the placement problem, for a given set of sensor and actuators, a subset is found which has controllability/observability properties close to the original set. Both problems are investigated in this chapter.

Considering the assignment problem, one shall notice that not every controllability and observability properties can be acquired with a given set of actuator and sensor location. However, for flexible structures in the balanced coordinates the situation is more approachable: the solution of the grammian assignment problem exists, or at least an approximate solution exists. The uniform assigning is a special but important problem considered. In this case the same degree of controllability and/or observability of each state is obtained, and the solution always exists.

The placement problem is also solved in the balanced coordinates, using the previously derived properties of the balanced flexible structures. In this case a comparatively simple methodology of choice of small subset of sensors and/or actuators from a large set of possible locations is derived.

5.1 Balanced Grammian Assignment

For the control system purposes, it is advantageous to have a tool for modifying or shaping a system controllability and observability properties. This can be done in two ways: by modifying the system properties, such as introducing a feedback loop, see Ref.[112], or by determining proper sensor and/or actuator configurations, see Refs.[32], and [35] (the configuration includes the actuator and sensor location, as well as the gains at each location). The latter method, called the grammian assignment problem, is addressed here. This approach has been found useful in control design (to avoid spillover) or in identification (placing actuators and sensors to recover the full-order system).

5.1.1 Problem Statement

Let the system be described by a stable system matrix A, and the location of actuators and/or sensors is not yet known (thus B and/or C matrices are not yet known). The following problem is considered. For a stable system and a positive definite matrix W find a state-space representation of a system such that its grammians are equal to W. It can be solved by finding B_1, C_1, and nonsingular transformation R, such that the grammians of (A_1, B_1, C_1) are equal to W, and $A_1 = R^{-1}AR$. Depending on what is to be determined (sensors, actuator, or both), the problem is divided into three separate problems.

Problem 5.1   Given A, B, find C_1 and the transformation R such that $W_{c1} = W_{o1} = W$ for the representation (A_1, B_1, C_1), where $A_1 = R^{-1}AR$, and $B_1 = R^{-1}B$.

Problem 5.2   Given A, C, find B_1 and the transformation R such that $W_{c1} = W_{o1} = W$ for the representation (A_1, B_1, C_1), where $A_1 = R^{-1}AR$, and $C_1 = CR$.

Problem 5.3   Given A, find B_1 and C_1 and the transformation R such that $W_{c1} = W_{o1} = W$, for the representation (A_1, B_1, C_1), where $A_1 = R^{-1}AR$.

Note that the matrices B_1 and C_1 include not only the actuator and sensor location, but also the gain at each location (for the location-only problem the entries of B_1 and C_1 would be either 1 or 0).

The solution of the problems may or may not exist, since not every positive definite grammian can be obtained through the sensor or actuator placement. A matrix W, for which the solution exists is an assignable one. A symmetric positive definite matrix W is c-assignable with respect to (A,C) if there exist a nonsingular transformation R and input matrix B such that the controllability grammian of $(R^{-1}AR, R^{-1}B)$ is equal to W. A symmetric positive definite matrix W is o-assignable with respect to (A,B) if there exists a nonsingular transformation R and output matrix C such that the observability grammian of $(R^{-1}AR, CR)$ is equal to W. A symmetric positive definite matrix W is assignable with respect to A if there exists a nonsingular transformation R and matrices B, C such that the controllability and observability grammians of $(R^{-1}AR, R^{-1}B, CR)$ are equal to W.

A symmetric positive semi-definite matrix W is c-assignable if N_c is positive semi-definite. It is o-assignable if N_o is positive semi-definite, where

$$N_c = -AW - WA^T, \qquad N_o = -A^TW - WA. \tag{5.1}$$

If it is c- and o-assignable, it is assignable.

5.1.2 Assignment Algorithms

Assuming c- and/or o-assignable W, the following algorithms solve problems 5.1, 5.2, and 5.3 respectively.

Algorithm 5.1 For given A, B, and matrix W o-assignable and positive definite:

1. Determine W_c from the Lyapunov equation (3.3).
2. Find P_c and P from the decomposition of W_c and W: $W_c = P_c P_c^T$, $W = PP^T$.
3. Find A_1, B_1: $A_1 = R^{-1}AR$, $B_1 = R^{-1}B$, $R = P_c P^{-1}$.
4. Define $N_o = -A_1^T W - WA_1$, and determine C_1 from the decomposition of N_o: $N_o = C_1^T C_1$ (note that N_o is positive semi-definite for W o-assignable).

In order to prove the algorithm, note from step 4 that $A_1^T W + WA_1 + C_1^T C_1 = 0$, i.e., W is the assigned observability grammian. For A_1, B_1 from step 3 one obtains $A_1 W + WA_1^T + B_1 B_1^T = 0$, hence W is the assigned controllability grammian.

A similar algorithm is obtained for determination of actuator locations.

Algorithm 5.2 For given A, C, and matrix W c-assignable and positive definite:

1. Determine W_o from the Lyapunov equation (3.3).
2. Find P_o and P such that $W_o = P_o^T P_o$, $W = P^T P$.
3. Find A_1, C_1: $A_1 = R^{-1}AR$, $C_1 = CR$, $R = P_o^{-1}P$.
4. Define $N_c = -A_1 W - WA_1^T$, and determine B_1 from the decomposition of N_c: $N_c = B_1 B_1^T$ (note that N_c is positive semi-definite for W c-assignable).

Problem 5.3 is solved by the following algorithm:

Algorithm 5.3 For W assignable and positive definite, the matrices B_1 and C_1 obtained from the decomposition of N_c and N_o: $N_c = B_1 B_1^T$, $N_o = C_1^T C_1$ are the solution of Problem 5.3.

5.1.3 Grammian Assignment for Flexible Structures

As a remainder, note that the balanced grammians for flexible structures have the following form

$$W_c = W_o = diag(\gamma_1^2, \ \gamma_1^2 + \varepsilon_1, \ \gamma_2^2, \ \gamma_2^2 + \varepsilon_2, ..., \ \gamma_{n2}^2, \ \gamma_{n2}^2 + \varepsilon_{n2}) \qquad (5.2)$$

where $|\varepsilon_i| \ll \gamma_i$, $i=1,...,n_2$ (i.e., each pair of states is described by two almost equal Hankel singular values). Thus the following is a sufficient condition of assignability of structures for A_i modal form 1.

Let $W_i = \gamma_i^2 I_2$, $\gamma_i \geq 0$, $A = diag(A_i)$, where A_i is in modal form 1, then $W = diag(W_i)$, $i=1,...,n_2$ is assignable. It is proved by noting that $N_c = N_o = diag(N_i)$, where $N_i = 4\zeta_i \omega_i \gamma_i^2 \ diag(0,1)$, hence N_c and N_o are positive-definite, and W is assignable. This condition is seldom met, but is useful for approximate assignment, defined as follows.

Let $\lambda^- = [\lambda_1^-,...,\lambda_p^-]$ be a vector of all negative eigenvalues of a symmetric matrix N, and $\lambda^+ = [\lambda_1^+,...,\lambda_q^+]$ be a vector of all non-negative eigenvalues of N. The matrix N is almost positive semi-definite if $\|\lambda^-\| \ll \|\lambda^+\|$, where $\|.\|$ is the Euclidean norm. A symmetric positive semi-definite matrix W is approximately c-assignable with respect to A if N_c is almost positive semi-definite; it is approximately o-assignable with respect to A if N_o is almost positive semi-definite. If it is approximately c- and o-assignable, it is approximately assignable.

The error ε of the approximate assignment

$$\varepsilon = \frac{\| W_{co} - W \|}{\| W \|}, \qquad W_{co} = (W_c W_o)^{1/2} \qquad (5.3)$$

quantifies the accuracy of the assignment.

As was shown in the previous chapters, the modal representation of a structure is almost balanced, with the grammians diagonally dominant. Thus, it follows from the previous condition that the grammian

$$W = diag(\gamma_i^2 I_2), \quad i=1,\dots,n_2 \qquad (5.4)$$

is approximately assigned.

Consider now an assignment of each mode separately, i.e., assign a grammian

$$W_i = diag(0,0,\dots,\gamma_i^2 I_2,\dots,0) \qquad (5.5)$$

Each mode can be approximately assigned, and the combined assignment is a sum of assignments of each individual mode. Denote $B^T = [B_1^T, B_2^T, \dots, B_{n2}^T]$, $C = [C_1, C_1, \dots, C_{n2}]$, where B_i is a two-row block of B, C_i is a two-column block of C, and denote also

$$B_{oi} = \begin{bmatrix} 0 \\ 0 \\ \dots \\ B_i \\ \dots \\ 0 \end{bmatrix}, \qquad C_{oi} = [0,0,\dots,C_i,\dots,0] \qquad (5.6)$$

and W, W_i as in Eqs.(5.4) and (5.5) respectively. If (A,B,C) is approximately assigned with W, then (A,B_{oi},C_{oi}) is approximately assigned with W_i. Conversely, if (A,B_{oi},C_{oi}), $i=1,\dots,n_2$ are

112

approximately assigned with W_i, then (A,B,C) is approximately assigned with W. The proof follows from the modal representation in form 1 or 2, for which Eq.(3.60) holds, hence

$$B_{oi}B_{oi}^T \cong C_{oi}^T C_{oi} \cong -W_i(A_i^T + A_i) \cong -(A_i^T + A_i)W_i \tag{5.7}$$

is true, i.e., W_i is approximately assigned. To prove the second part, note that $B = \sum_{i=1}^{n2} B_{oi}$, $C = \sum_{i=1}^{n2} C_{oi}$. It follows from Eq.(5.7) that the components of B are almost orthogonal, $B_{oi}B_{oj}^T \cong 0$, $i \neq j$, as well as the components of C, $C_{oi}^T C_{oj} \cong 0$, $i \neq j$, hence $W = \sum_{i=1}^{n2} W_i$. The latter fact shows that W is approximately assigned if W_i, $i=1,..,n_2$ are approximately assigned.

The above property allows one to assign each mode separately, and to obtain results for a combination of modes. Conversely, by determining the assignment for all modes, one can approximately assign a single mode by specifying B, C as in Eq.(5.6), as will be illustrated in the example below.

The approximate assignment can be achieved by using modal parameters ζ_i, ω_i, since it follows from Eq.(3.60) that

$$B_i B_i^T \cong C_i^T C_i \cong -\gamma_i^2(A_i^T + A_i) = diag(0, \quad 4\zeta_i \omega_i \gamma_i^2). \tag{5.8}$$

Therefore, one obtains

$$B_i^T = C_i = \begin{cases} 2\zeta_i \omega_i \gamma_i^2 [1 \quad 1] & \text{for the modal form 1} \\ 2\zeta_i \omega_i \gamma_i^2 [0 \quad 2] & \text{for the modal form 2} \end{cases} \tag{5.9}$$

which is the approximate assignment for collocated inputs and outputs.

113

If the approximation error is too high to be accepted, the following iterative procedure reduces the error.

Algorithm 5.4 Given approximation accuracy ε_o, positive semi-definite W, and system matrices A and B, assume C. Denote this triple $(A^{(1)}, B^{(1)}, C^{(1)})$ and its balanced grammian $W^{(1)}$. Determine P: $W=PP^T$. For $i=1,2,\ldots$

1. Determine $P_c^{(i)}$ from the decomposition of $W^{(1)}$: $W^{(1)}=P_c^{(1)}P_c^{(1)T}$.

2. Determine $\qquad A^{(i+1)}, \qquad\qquad B^{(i+1)}$: $\qquad\qquad A^{(i+1)}=R^{(i)-1}A^{(i)}R^{(i)}$, $B^{(i+1)}=R^{(i)-1}B^{(i)}$, $R^{(i)}=P_c^{(i)}P^{(i)-1}$.

3. Determine $N^{(i)}$: $N^{(i)}=-A^{(i+1)T}W-WA^{(i+1)}$.

4. Decompose $\quad N^{(i)}$: $\quad N^{(i)}=U^{(i)}\Lambda^{(i)}U^{(i)T}$, $\quad$ where $\quad \Lambda^{(i)}=diag(\lambda_k^{(i)})$, $k=1,\ldots,n$, and $\lambda_k^{(i)}$ is the kth eigenvalue of $N^{(i)}$.

5. Find the positive semi-definite replacement $N_p^{(i)}$ of $N^{(i)}$ by setting all negative eigenvalues of $N^{(i)}$ to zero: $N_p^{(i)}=U^{(i)}\Lambda_p^{(i)}U^{(i)}$, where $\Lambda_p^{(i)}=0.5(\Lambda^{(i)}+|\Lambda^{(i)}|)$.

6. Decompose $N_p^{(i)}$: $N_p^{(i)}=Q^{(i)T}Q^{(i)}$, $C^{(i+1)}=Q^{(i)}$.

7. Balance $(A^{(i+1)},B^{(i+1)},C^{(i+1)})$, obtaining $W^{(i+1)}$.

8. Check convergence, determining $\varepsilon^{(i+1)}$: $\varepsilon^{(i+1)}=\|W-W^{(i+1)}\|/\|W\|$. If $\varepsilon^{(i+1)}<\varepsilon_o$ then stop, otherwise set $i=i+1$ and go to 1.

The grammian assignment approach is useful in the system identification, when, for example one would like to identify only one specific mode, a group of natural modes, or all modes of a structure. The assignment procedure will locate the sensors or actuators such that the required modes are excited, while the remaining ones are at rest.

Example 5.1 For a system in Fig.2.1 with the masses $m_1=m_2=m_3=1$, stiffness $k_1=k_4=100$, $k_2=50$, $k_3=10$, and the damping matrix D is proportional to the stiffness matrix K, i.e., $D=0.001K$. Two input forces are applied, such that the matrix B is as follows

$$B^T=\begin{bmatrix} 0 & 0 & 0 & 0.1 & 0.2 & -0.5 \\ 0 & 0 & 0 & -.02 & -0.3 & 0.1 \end{bmatrix}$$

114

The grammian $W=diag(10, 10, 7, 7, 1, 1)$ is assigned by determining the matrix C of sensor locations.

Using Algorithm 5.1, one obtains the triple (A_1, B_1, C_1)

$$A_1 = \begin{bmatrix} -0.0363 & 6.0463 & -0.0499 & -0.0019 & 0.0063 & 0.0004 \\ -6.0526 & -0.0003 & 0.0030 & 0.0003 & -0.0006 & 0.0000 \\ 0.1042 & 0.0042 & -0.1105 & -10.523 & 0.0351 & 0.0024 \\ -0.0034 & -0.0006 & 10.5354 & -0.0004 & 0.0020 & 0.0001 \\ 0.0318 & 0.0027 & 0.0109 & -0.0001 & -0.1718 & -13.124 \\ -0.0022 & -0.0002 & -0.0016 & 0.0000 & 13.1468 & -0.0008 \end{bmatrix}$$

$$B_1^T = \begin{bmatrix} 0.3620 & 0.0349 & -1.1979 & 0.0640 & 0.2423 & -0.0170 \\ -0.7715 & -0.0649 & 0.3354 & -0.0315 & 0.5337 & -0.0354 \end{bmatrix}$$

$$C_1 = \begin{bmatrix} 0.2896 & 0.0561 & -1.2308 & 0.0709 & 0.2188 & 0.0078 \\ -0.7874 & -0.0559 & -0.1387 & -0.0361 & 0.2881 & -0.0040 \\ 0.1495 & 0.0149 & 0.1152 & -0.0405 & 0.4612 & -0.0502 \\ -0.0039 & 0.0370 & 0.0062 & 0.0655 & 0.0082 & 0.0365 \\ 0.0008 & -0.0338 & 0.0007 & 0.0377 & 0.0003 & -0.0335 \\ 0.0030 & -0.0289 & 0.0001 & -0.0001 & 0.0031 & 0.0291 \end{bmatrix}$$

For this representation the grammians are $W_c = W_o = diag(10.0610, 10.0609, 7.0391, 7.0390, 1.0131, 1.0130)$; i.e., it is approximately assigned, with the approximation error $\varepsilon = 0.0030$. This assignment is improved iteratively with Algorithm 5.4, obtaining the assignment error as in Fig.5.1. It shows that with 21 iterations the error improved from 0.003 to 0.0007.

Next, the assignment for $W_1 = diag(0,0,1,1,0,0)$ and $W_2 = diag(0,0,1,1,1,1)$ is examined. In the first case only the second mode is excited and observed, while in the second case the second and the third modes are excited and observed. As mentioned previously, the matrix A_1 as shown above assigns W_1 and W_2, with the matrices B_{o1}, C_{o1}, and B_{o2}, C_{o2} obtained from B_1, C_1 (given above) by setting to zero or scaling appropriate rows of B_1 and columns of C_1. In the first case $(1,2,5,6)$ columns of C_1 and rows of B_1 are set to zero, and $(3,4)$

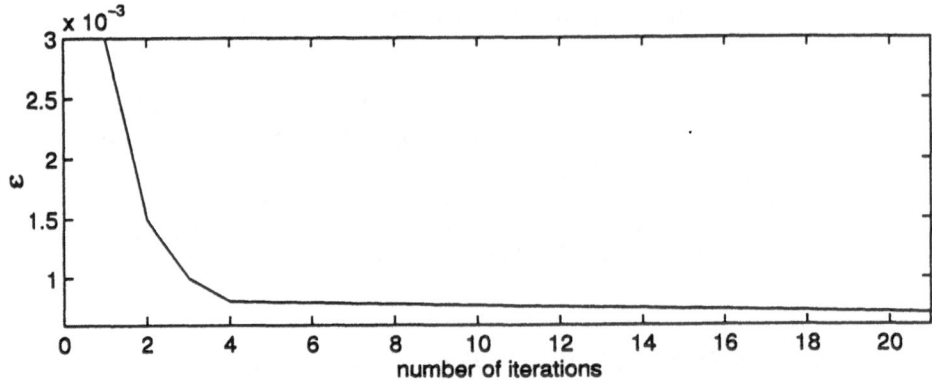

Figure 5.1: Iterative improvement of the assignment error

columns of C_1 and rows of B_1 remain unchanged. In the second case (1,2) columns of C_1 and rows of B_1 are set to zero, and (3,4,5,6) columns of C_1 and rows of B_1 remain unchanged. Plots of the magnitude of transfer function for the first case are shown in Fig.5.2a, and for the second case in Fig.5.2b. The figures show clearly that in the first case, only the first mode is excited and observed, and in the second case, the first and the third modes are excited and observed. The assignment error is $\varepsilon = 0.0197$ in the first case, and $\varepsilon = 0.0419$ in the second case.

Example 5.2 A flexible truss shown in Fig.2.2 is assigned with $W=I$ and $W=diag(1,1,0,0,...,0)$. In the first case all its modes are equally controllable and observable, in the second case only its first mode is controllable and observable. The results of the assignment are shown in Fig.5.3a,b. For modes equally controllable and observable the related resonant peaks are equal to two. For instance, the resonant peak for the only controllable and observable mode is equal to two, while all other resonances have been eliminated.

116

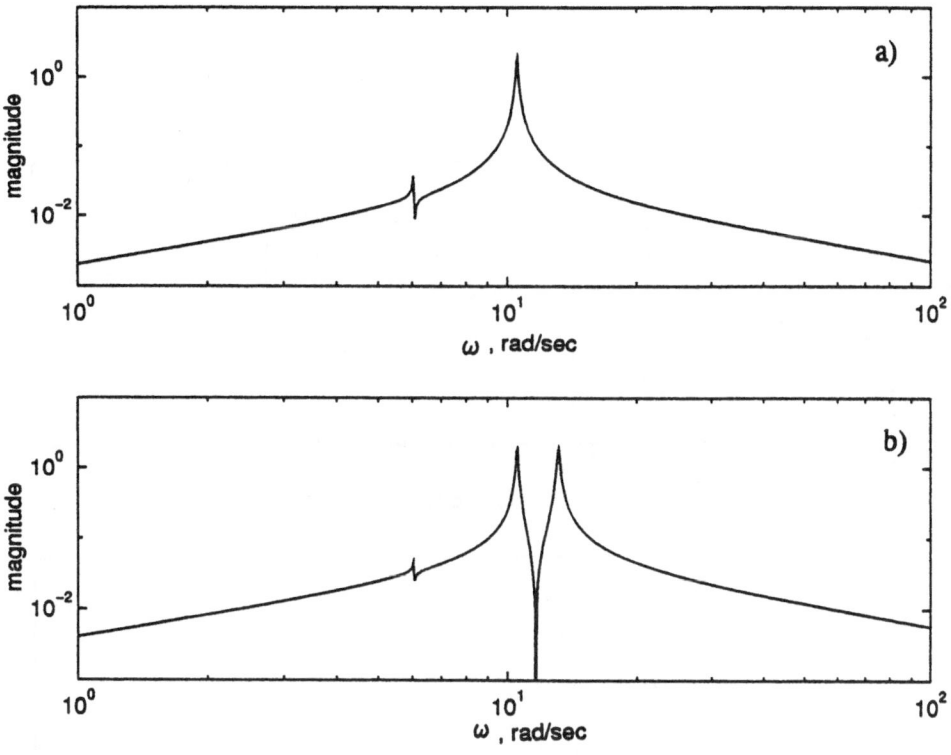

Figure 5.2: Magnitude of transfer function (a) for the first case assignment, (b) for the second case assignment

5.1.4 Uniformly Balanced Systems and Structures

A system with all its states equally controllable and observable is called uniformly balanced. All its Hankel singular values are equal, so that the grammians have the form $W_c = W_o = \gamma^2 I_n$, where $\gamma > 0$. For a stable A the uniformly balanced system is always assignable. Indeed, in this case $BB^T = C^T C = -\gamma^2 (A + A^T)$ and $-A - A^T$ is positive-definite, hence one can find B and C by decomposition of $-\gamma^2 (A + A^T)$. The uniformly balanced system is determined as follows.

117

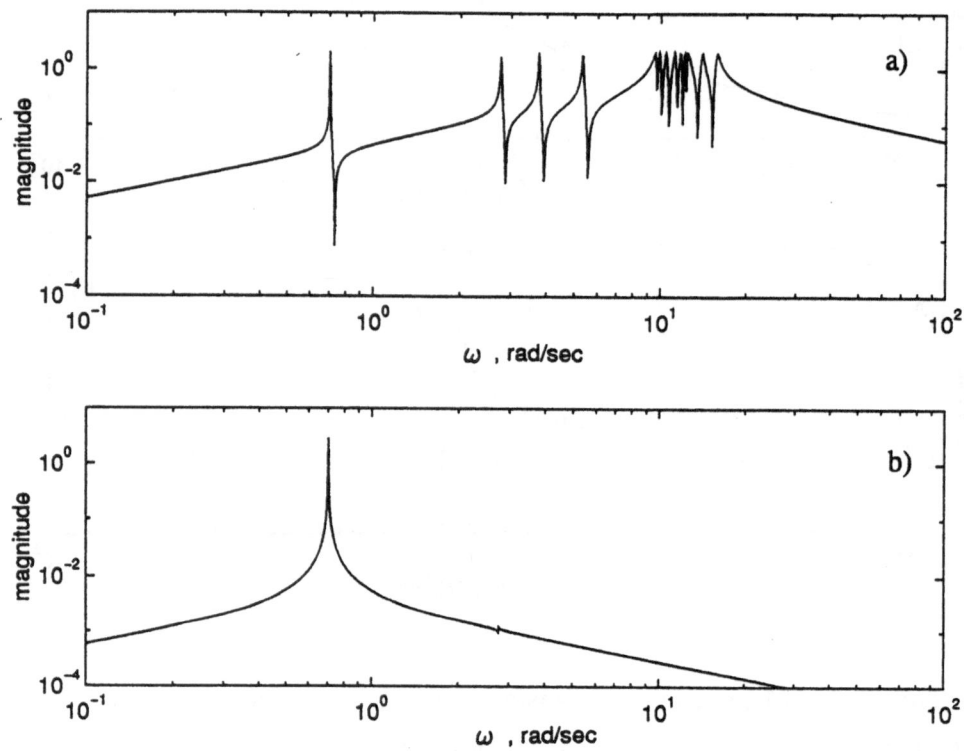

Figure 5.3: Magnitude of the truss transfer function: (a) all modes equally controllable and observable, (b) only the first mode controllable and observable

Algorithm 5.5   Given A, B, the grammian $W=\gamma^2 I_n$ is assigned with $C=B^T W_c^{-1}\gamma^2$, where W_c the controllability grammian of (A,B).
Algorithm 5.6   Given A, C, the grammian $W=\gamma^2 I_n$ is assigned with $B=W_o^{-1}C^T\gamma^2$, where W_o is the observability grammian of (A,C).

Properties of the uniformly balanced systems are obtained from the following auxiliary attributes. Let $G(j\omega)=C(j\omega I-A)^{-1}B$ be a transfer function of the triplet (A,B,C), and $s=j\omega$ be not a pole of an element of $G(s)$. For $p \leq m$ let the $p \times m$ matrix $U_1(\omega)$ have p orthonormal rows, i.e., $U_1(\omega)U_1^*(\omega)=I_p$, and for $p \geq m$ let the $p \times m$ matrix $U_2(\omega)$ have m orthonormal columns; i.e., $U_2(\omega)U_2^*(\omega)=I_m$. If $rank(U_1(\omega))=rank(G(j\omega))=p$ for all real ω, and G satisfies the equation

$$GU_1+U_1^*G^* = \frac{GG^*}{\gamma^2} \tag{5.10}$$

or if $rank(U_2(\omega))=rank(G(j\omega))=m$, and G satisfies the equation

$$U_2G+G^*U_2^* = \frac{G^*G}{\gamma^2} \tag{5.11}$$

then the system is stable.

In order to prove it, introduce $G_1=GU_1$ and $G_2=U_2G$. It follows from Eq.(5.10) that G_1 is positive real, and from Eq.(5.11) that G_2 is positive real, hence both are stable. But (A,BU_1,C) is the state-space representation of G_1, and (A,B,U_2C) is the state-space representation of G_2. The representations G, G_1, and G_2 include stable matrix A, hence G, G_1, and G_2 are stable.

Now we can claim that the transfer function $G(j\omega)=C(j\omega I-A)^{-1}B$ of a uniformly balanced system (A,B,C) has the following properties:

(a) G is stable,
(b) $\|G\|_u \leq 2\gamma^2$,
(c) If the system is square (the number of inputs is equal to the number of outputs, m) then G is positive real, and $G_o=\gamma^{-2}G-I_m$ is all pass.

119

Above, $\|.\|_u$ is a unitarily invariant norm, e.g., $\|G\|_\infty = \sup_\omega$
$\bar{\sigma}(G(j\omega))$, or $\|G\|_2^2 = \frac{1}{2\pi} \int_\infty^\infty tr(GG^*)d\omega$, and $\bar{\sigma}(G)$ is the largest singular value of G.

In order to prove (a), note that G satisfies Eqs. (5.10) and (5.11), and thus is stable. To prove (b), note that from the left-hand side of Eq.(5.10) it follows that $\|GU + U^*G^*\|_u \le 2\|GU\|_u = 2\|G\|_u$. From the right-hand side of Eq.(5.10) one obtains $\|GG^*\|_u/\gamma^2 = \|G\|_u^2/\gamma^2$, thus $\|G\|_u^2/\gamma^2 \le 2\|G\|_u$, or $\|G\|_u \le 2\gamma^2$. The positive realness of the square uniformly balanced system follows from Eq.(5.10) and the definition of positive-realness of a transfer function, see [2]. Introducing $G = \gamma^2(G_o + I_m)$ to Eqs.(5.10) or (5.11) one obtains $G_o^*G_o = G_oG_o^* = I_m$, hence G_o is all-pass.

The plot of the transfer function for a single-input single-output uniformly balanced system is a circle of radius γ^2, and center $(\gamma^2, 0)$.

The uniformly balanced representation is useful in the system identification, since in this representation the locations of sensors and/or actuators are such that the all states (modes) are equally excited and observed, hence they are equally "treated" when tested.

Example 5.3 The uniformly balanced representation with $\gamma = 1$ is determined for A as in Example 5.1, and $B^T = [0\ 0\ 0\ 1\ 0\ 0]$. From Algorithm 5.5 we arrive at the triple (A, B, C)

$$A = \begin{bmatrix} 0.0000 & 6.0485 & 0.1659 & 0.0085 & 0.0003 & 0.0484 \\ -6.0510 & -0.0377 & -0.0913 & -0.0296 & 0.0596 & -0.0589 \\ -0.1716 & -0.0755 & -0.1848 & -13.142 & 0.2270 & -0.1477 \\ -0.0090 & 0.0163 & 13.1121 & -0.0012 & 0.0305 & -0.0819 \\ -0.0005 & -0.0650 & -0.2389 & -0.0314 & -0.0002 & 10.5323 \\ -0.0525 & -0.0615 & -0.1189 & 0.0606 & -10.5409 & -0.0962 \end{bmatrix}$$

120

$$B^{\mathrm{T}} = C = [0.0094 \quad 0.2744 \quad 0.6079 \; 0.0485 \quad 0.0195 \quad 0.4387]$$

which is uniformly balanced. The plot of the transfer function of (A,B,C) in Fig.5.4 shows that that the property (b) is satisfied.

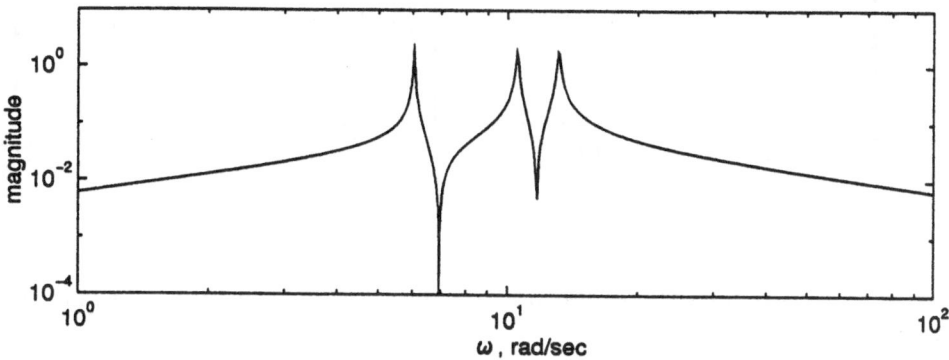

Figure 5.4: Magnitude of transfer function of the uniformly balanced system for $\gamma = 1$

The uniformly balanced truss structure has been presented in Example 5.2. The Deep Space Network antenna model is uniformly balanced in the following example.

Example 5.4 The previously presented model of the Deep Space Network antenna is uniformly balanced. The elevation rate command is selected as a single input to the antenna. The Hankel singular values of all states are set equal to one, except for states with poles at zero - their Hankel singular values are infinitely large. These states are removed from the model for clarity of presentation. The results are presented in Fig.5.5, where the magnitude of the transfer function of the uniformly balanced antenna is presented; the magnitude, according to the property (b), is less than two.

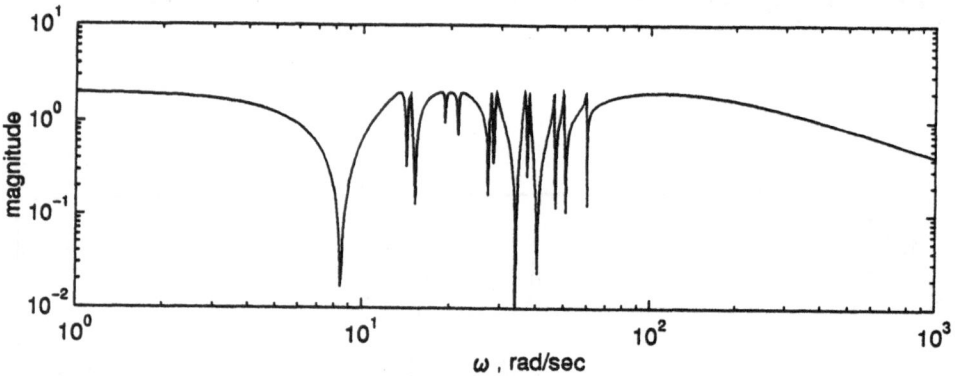

Figure 5.5: Magnitude of transfer function of the uniformly balanced antenna control system

5.2 Actuator and Sensor Placement

The importance of actuator and sensor placement problems is underscored in many investigations and contributions, only part of which we have included in our references (see, for example, Refs.[1], [7], [19], [20], [52], [63], [68], [69], [73], [74], [75], [[79], [81], [82], [85], [86], [99], [104], and [109]). Here, an approach based on the balanced representation of flexible structures is presented. The Hankel singular values indicate the importance of the sensors and the actuator in the balanced coordinates, thus they can be used for the choice of inputs and also outputs such that controllability and observability properties of the system are maximized.

122

5.2.1 Additive Property of the Hankel Singular Values

In this chapter the sensor and/or actuator locations are determined to maximize the controllability and observability properties of a system. These properties are quantified by the Hankel singular values, subsequently used as indicators of the validity of sensor and actuator locations. This option has not yet been investigated thoroughly, since in a general case, it is difficult to derive an explicit relationship between the sensor/actuator locations and the Hankel singular values. Fortunately, this relationship can be obtained for flexible structures.

Let $\mathcal{P}$ and $\mathcal{Q}$ be the sets of the candidate actuator and sensor locations, respectively. These are chosen in advance to be all allowable locations of actuators, the population of which is P, and all allowable locations of sensors, population Q. The placement of p actuators within the given $\mathcal{P}$ actuator candidate locations, and the placement q sensors within the given Q sensor candidate locations, is considered. Of course, the number of candidate locations must be larger than the number of actuators or sensors, i.e., $p<P$ and $q<Q$. In order to analyze this case the input B and the output C matrices must consist of all candidate locations, i.e., of P columns and Q rows, respectively

$$B=[B_1,\dots,B_P], \qquad C=\begin{bmatrix} C_1 \\ \cdots \\ \cdots \\ C_Q \end{bmatrix} \qquad (5.12)$$

These matrices have the following property

$$BB^T = \sum_{i=1}^{P} B_i B_i^T, \qquad C^T C = \sum_{j=1}^{Q} C_j^T C_j \qquad (5.13)$$

therefore the controllability (W_c) and observability (W_o) grammians (as the solutions of the Lyapunov equations (3.3) for BB^T and C^TC as in

123

Eq.(5.13)) are a sum of grammians for each individual sensor and actuator

$$W_c = \sum_{i=1}^{P} W_{ci}, \qquad W_o = \sum_{j=1}^{Q} W_{oj} \qquad (5.14)$$

In the above equation W_{ci} and W_{oj} are the controllability and observability grammians for the ith input and jth output, respectively. The Hankel singular values are obtained from the eigenvalues of the product of the grammians

$$\gamma_k^4 = \lambda_k(W_c W_o) \qquad (5.15)$$

where $\lambda_k(X)$ denotes the kth eigenvalue of a square matrix X. Introducing Eq.(5.14) to Eq.(5.15), one obtains

$$\gamma_k^4 = \lambda_k \left(\sum_{i=1}^{P} \sum_{j=1}^{Q} W_{ci} W_{oj} \right) \qquad (5.16)$$

Unfortunately, this equation cannot be explicitly solved in a general case. However, for flexible structures the balanced representation is almost independent of actuator and sensor location, see Eq.(3.54). Thus if a flexible structure is balanced, one obtains

$$W_{ci} W_{oj} \cong \Gamma^4(i,j), \qquad \Gamma(i,j) = diag(\gamma_1(i,j),\dots,\gamma_{n2}(i,j)) \qquad (5.17)$$

where $\Gamma(i,j)$ is the matrix of the Hankel singular values for the ith actuator and the jth sensor location. Introducing Eq.(5.17) to Eq.(5.16), we arrive at

$$\gamma_k^4 \cong \sum_{i=1}^{P} \sum_{j=1}^{Q} \gamma_k^4(i,j), \qquad k=1,\dots,n_2 \qquad (5.18)$$

This additive property of flexible structures, says that the fourth power of the Hankel singular values of a system with multiple sensors and actuators is the approximate sum of the fourth powers of the Hankel singular values of the systems with a single sensor and a single actuator. The importance of Eq.(5.18) lies in the decomposition of the problem of the actuator and sensor placement into a sum of separate placement problems for each individual sensor and actuator. Each sensor and actuator is individually evaluated for each balanced mode through the parameter $\gamma_k^4(i,j)$, where $\gamma_k(i,j)$ is the Hankel singular value of the kth balanced mode for the ith actuator and jth sensor.

5.2.2 Placement Matrix and Placement Indices

Three separate problems can be distinguished: actuator placement only, sensor placement only, and joint actuator and sensor placement. Due to the property of Eq.(5.18), all three are similar. Thus, for simplicity of presentation, the problem of actuator placement is considered only. In this case the equation (5.18) simplifies to

$$\gamma_k^4 \cong \sum_{i=1}^{P} \gamma_k^4(i), \qquad k=1,\dots,n_2 \qquad (5.19)$$

where $\gamma_k(i)$ is the kth Hankel singular value for the ith actuator location, and γ_k is the kth Hankel singular value for all P locations.

Next, the placement index of the ith actuator is defined. This index should reflect the design purposes, and it can be defined as the weighted sum over all modes

$$\gamma_a^4(i) = \sum_{k=1}^{n2} w_k^2 \gamma_k^4(i), \qquad i=1,\dots,P \qquad (5.20)$$

125

where $w_k \geq 0$ is the weight assigned to the kth mode. The weight reflects the importance of the mode in the controller design. If there is no other clue, it is assumed uniform, that is, $w_k = 1$, for $k = 1, \ldots, n_2$. The weight matrix is defined as $W = diag(w_k)$.

In application, it is convenient to represent the placement indices in the matrix form. First, the matrix G is introduced

$$
G = \begin{bmatrix}
\gamma_1^2(1) & \gamma_1^2(2) & \ldots & \gamma_1^2(i) & \ldots & \gamma_1^2(P) \\
\gamma_2^2(1) & \gamma_2^2(2) & \ldots & \gamma_2^2(i) & \ldots & \gamma_2^2(P) \\
\ldots & \ldots & \ldots & \ldots & \ldots & \ldots \\
\gamma_k^2(1) & \gamma_k^2(2) & \ldots & \gamma_k^2(i) & \ldots & \gamma_k^2(P) \\
\ldots & \ldots & \ldots & \ldots & \ldots & \ldots \\
\gamma_{n2}^2(1) & \gamma_{n2}^2(2) & \ldots & \gamma_{n2}^2(i) & \ldots & \gamma_{n2}^2(P)
\end{bmatrix}
\tag{5.21}
$$

By weighting and normalizing it, one obtains the placement matrix Σ, defined as follows

$$
\Sigma = \frac{WG}{\|WG\|}
\tag{5.22}
$$

where $\|.\|$ is a Frobenius norm, i.e., $\|WG\| = \sqrt{tr(WGG^T W^T)}$. The placement matrix has a form

$$
\Sigma = \begin{bmatrix}
\sigma_{11} & \sigma_{12} & \ldots & \sigma_{1i} & \ldots & \sigma_{1P} \\
\sigma_{21} & \sigma_{22} & \ldots & \sigma_{2i} & \ldots & \sigma_{2P} \\
\ldots & \ldots & \ldots & \ldots & \ldots & \ldots \\
\sigma_{k1} & \sigma_{k2} & \ldots & \sigma_{ki} & \ldots & \sigma_{kP} \\
\ldots & \ldots & \ldots & \ldots & \ldots & \ldots \\
\sigma_{n_2 1} & \sigma_{n_2 2} & \ldots & \sigma_{n_2 i} & \ldots & \sigma_{n_2 P}
\end{bmatrix} \leftarrow k\text{th mode}
\tag{5.23}
$$

$$
\uparrow
$$
ith actuator

and its $k i$th entry

$$\sigma_{ki} = \frac{w_k \gamma_k^2(i)}{\|WG\|} \qquad (5.24)$$

is the placement index that evaluates the ith actuator at the kth mode.

The placement matrix is a convenient tool for evaluation of each actuator. It gives an immediate insight into the placement properties of each (say ith) actuator, since the placement index of the ith actuator is determined as the root-mean-square (rms) sum of the ith column of Σ. The vector of the actuator placement indices, $\sigma_a = [\sigma_{a1}, \sigma_{a2}, ..., \sigma_{ap}]^T$, can be obtained from the placement matrix as

$$\sigma_a = sqrt(diag(\Sigma^T\Sigma)), \qquad (5.25a)$$

where $diag(X)$ is a vector formed from the elements of the main diagonal of the square matrix X, and $sqrt(x)$ is the square root of the elements of the vector x. The ith entry of σ_a

$$\sigma_{ai} = \sqrt{\sum_{k=1}^{n_2} \sigma_{ki}^2} \leq 1, \qquad i=1,....,P, \qquad (5.25b)$$

is the placement index of the ith actuator, and it is the rms sum of the ith actuator indexes over all modes.

Similarly, the vector of the modal indices, $\sigma_m = [\sigma_{m1}, \sigma_{m2},..., \sigma_{mn_2}]^T$, can be obtained from the placement matrix as

$$\sigma_m = sqrt(diag(\Sigma\Sigma^T)), \qquad (5.26a)$$

The kth entry of σ_m

$$\sigma_{mk} = \sqrt{\sum_{i=1}^{P} \sigma_{ki}^2} \leq 1, \qquad k=1,\ldots,n_2 \tag{5.26b}$$

is the index of the kth mode, and it is the rms sum of the kth mode indexes over all actuators.

The actuator placement index, σ_{ai}, is a non-negative contribution of the ith actuator at all modes to the controllability and observability properties of the structure. The modal index, σ_{mk}, is a non-negative contribution of the kth mode for all actuators to the controllability and observability properties of the structure.

From the above properties it follows that the index σ_{ai} characterizes the importance of the ith actuator, thus it serves as the actuator placement index. Namely, the actuators with small index σ_{ai} can be removed as the least significant ones. Note also that the balanced modal index σ_{mk} can also be used either as a placement index in cases when modes of the required controllability and observability level are sought, or as a reduction index. Indeed, it characterizes the significance of the kth balanced mode for the given locations of sensors and actuators. The controllability and observability of the least significant modes (those with the small index σ_{mk}) should either be enhanced by the reconfiguration of the actuators or sensors, or be eliminated, i.e., the system order is reduced.

5.2.3 Examples

Example 5.5 The truss from Fig.2.2 is considered. Its outputs are rates measured at vertical direction at nodes $n4$ and $n8$, respectively. The following actuators are considered: (1) force in the bar connecting node $n2$ and the basis, (2) force in the bar connecting node $n3$ and $n2$, (3) force in the bar connecting node $n4$ and $n3$, (4) force in the bar connecting node *n6* and the basis, (5) force in the bar connecting node *n7* and *n6*, and (6) force in the bar connecting node *n8* and *n7*. The modal weight is assumed uniform, thus $w_k=1$, for all modes. The task is to find the two inputs within the given six candidates with the best controllability and observability properties.

The controllability and observability properties of each actuator are characterized by the indices σ_{ai}, $i=1,...,6$. These indices are obtained from the Hankel singular values of each individual actuator, which contribute to the total Hankel singular values. This property is based on the approximate equality (5.18). First, the accuracy of the additive property is checked by computing the Hankel singular values of the system with all six inputs, and by determining the same Hankel singular values through Eq.(5.18). The results shown in Fig.5.6 confirm that (5.18) holds with satisfactory accuracy, and that, while discrepancies occurred for the small Hankel singular values, they are acceptable for the actuator placement purposes. The indices σ_{ai}, $i=1,...,6$ are given in Table 5.1, which indicates that input 1 (force in the bar connecting node *n2* and the basis) and input 4 (force in the bar connecting node *n6* and the basis) are the most appropriate for the actuator locations.

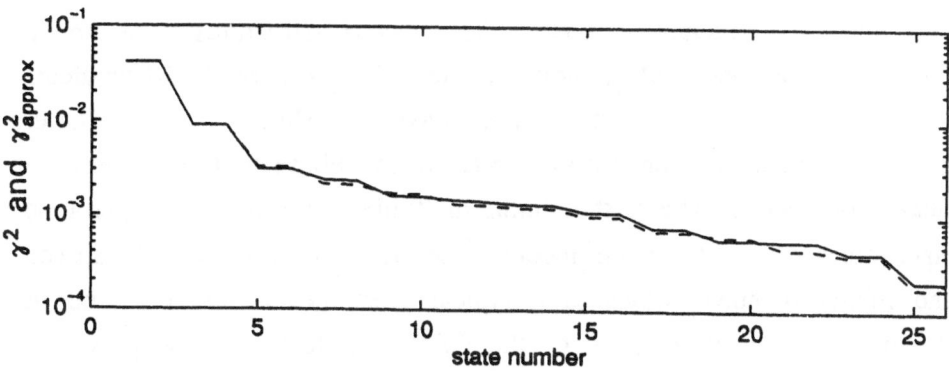

Figure 5.6: The accuracy of the Hankel singular value additive property for the first truss example

129

Table 5.1. Placement indices, σ_{ai}

Actuator	Example 5.5	Example 5.6
1	0.6042	0.0813
2	0.3304	0.3465
3	0.1245	0.0893
4	0.6231	0.6043
5	0.3267	0.0825
6	0.1235	0.3466
7	--	0.0903
8	--	0.6043

Example 5.6 The same truss with the same outputs is considered. The eight candidate actuator locations are given: (1) horizontal force at node $n3$, (2) vertical force at node $n3$, (3) horizontal force at node $n4$, (4) vertical force at node $n4$, (5) horizontal force at node $n7$, (6) vertical force at node $n7$, (7) horizontal force at node $n8$, and (8) vertical force at node $n8$. The task is to find the best two inputs within the candidate locations.

The actuator indices σ_{ai}, obtained from (5.24), are presented in Table 5.1. The results indicate that the locations 4 (vertical force at node $n4$), and 8 (vertical force at node $n8$) are the best choices.

Example 5.7 The actuator placement procedure is applied to an experimental structure called the Control-Structures Interaction Evolutionary Model (CEM), shown in Fig.5.7. A total of $P=50$ candidate locations for the air thrusters is selected and shown in this figure. The structural model consists of $n_2=12$ modes whose first six modes are suspension modes. The first column in Table 5.2 shows the open-loop eigenvalues of the structural modes. The frequencies are closely spaced and lightly damped, which is a typical phenomenon for this kind of structure. The main purpose of the CEM actuators is to suppress the suspension movement and structural vibrations due to external disturbances by using feedback control. Reference [8] gives a more detailed description of the structural model. Based on the set of

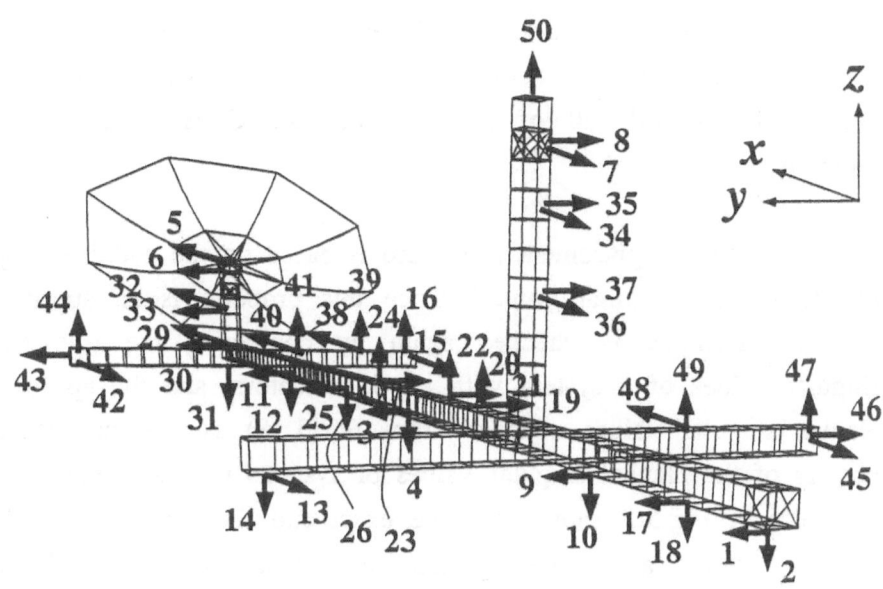

Figure 5.7: Actuator configuration for CEM

Table 5.2. Open-loop poles of CEM structure

Mode	Open-loop ω	ζ
1	0.930	0.037
2	0.941	0.001
3	0.973	0.050
4	4.623	0.005
5	4.752	0.004
6	5.614	0.002
7	9.410	0.002
8	14.54	0.001
9	15.29	0.006
10	15.54	0.001
11	16.27	0.006
12	18.29	0.014

actuators 1 to 8, the results of several control designs are given in Refs.[12], [75], [76], [77].

The placement of actuators depends on the sensor location. We consider three sensor locations. In the first case, sensors No. 9, 37, and 46 were used, all of them sensing the CEM dynamics in the y-direction.

The actuator placement procedure is based on the additive property of the Hankel singular values. Since this property is an approximate one, its accuracy is checked again by computing the true Hankel singular values of a system with all 50 actuators, and the approximate Hankel singular values obtained from Eq.(5.18) as the sum of fourth powers of the Hankel singular values of systems with a single actuator. The plot in Fig.5.8 shows that the exact and the approximate Hankel singular values are close enough to proceed with the placement procedure.

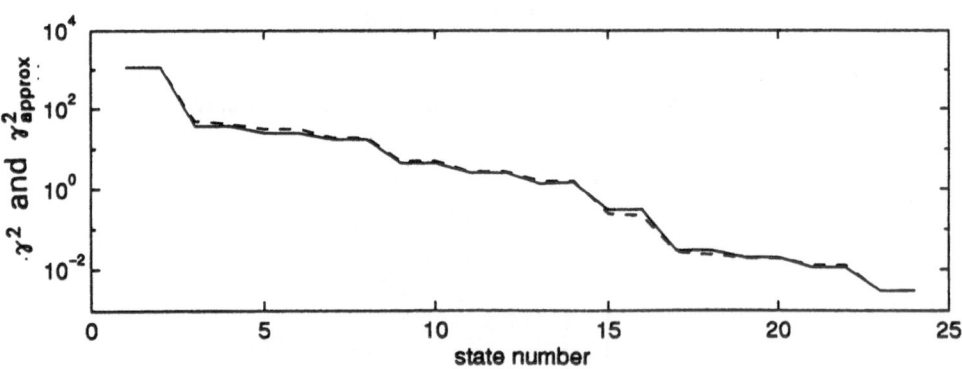

Figure 5.8: The accuracy of the Hankel singular value additive property for CEM

The modal indices σ_{mk} (for the uniform weights, $w_k=1$), for $k=1,...,24$, and the actuator placement indices σ_{ai}, for $i=1,...,50$ were determined and plotted in Fig.5.9a,b. The modal indices show that the second mode primarily participates in the output. The actuator indices show that actuators placed at locations 1, 3, 6, 8, 9, 11, 17, 19, 21, 23, 25, 27, 30, 33, 35, 37, 43, and 46 most influence the output. All of them act in y-direction, i.e., they are oriented in the same direction as the outputs.

Next, the output was configured in such a way that the first mode was sensed. The first mode consists primarily of the horizontal motion in the x-direction, see Fig.5.10a. The actuator placement indices obtained are plotted in Fig.5.11, showing that the locations 5, 7, 13, 15, 29, 32, 34, 36, 38, 40, 42, 45, 48 are good choices for actuator placement if one wants to excite the first mode.

Finally, the output is measured such that the last, the twelfth mode, was sensed. In this mode the antenna motion is dominant, as seen Fig.5.10b. The actuator placement indices are plotted in Fig.5.12a, showing that the locations 2, 4, 7, 18, 20, 22, 24, 26, 34 are the preferred placements for the actuators to excite the twelfth mode. In this case the location of actuators is not so obvious as in the previous cases. Although the mode consists mostly of the antenna motion, none of the preferred actuators is located on the antenna, or nearby it. They are located such that the antenna motion is excited by the rotation of the structure with respect to the y-axis.

In order to check whether the twelfth mode is excited, the modal indexes σ_{mk} were determined from Eq.(5.23) for $k=1,...,12$, and plotted in Fig.5.12b. The plot shows that mode 12 is indeed predominantly excited.

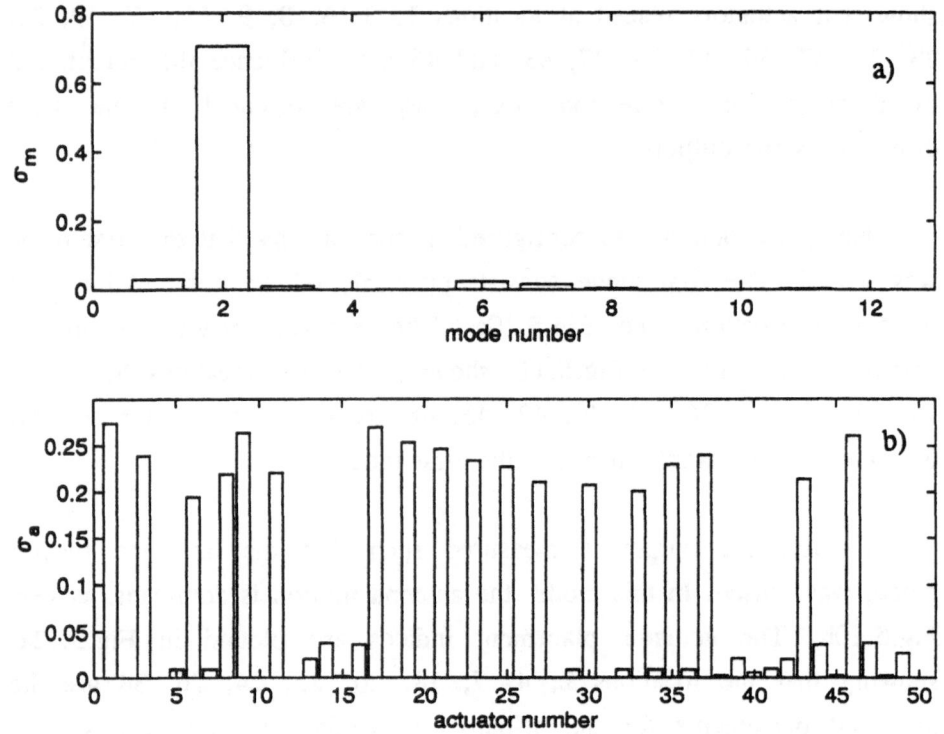

Figure 5.9: Modal indices (a), and actuator placement indices (b), for CEM, case 1

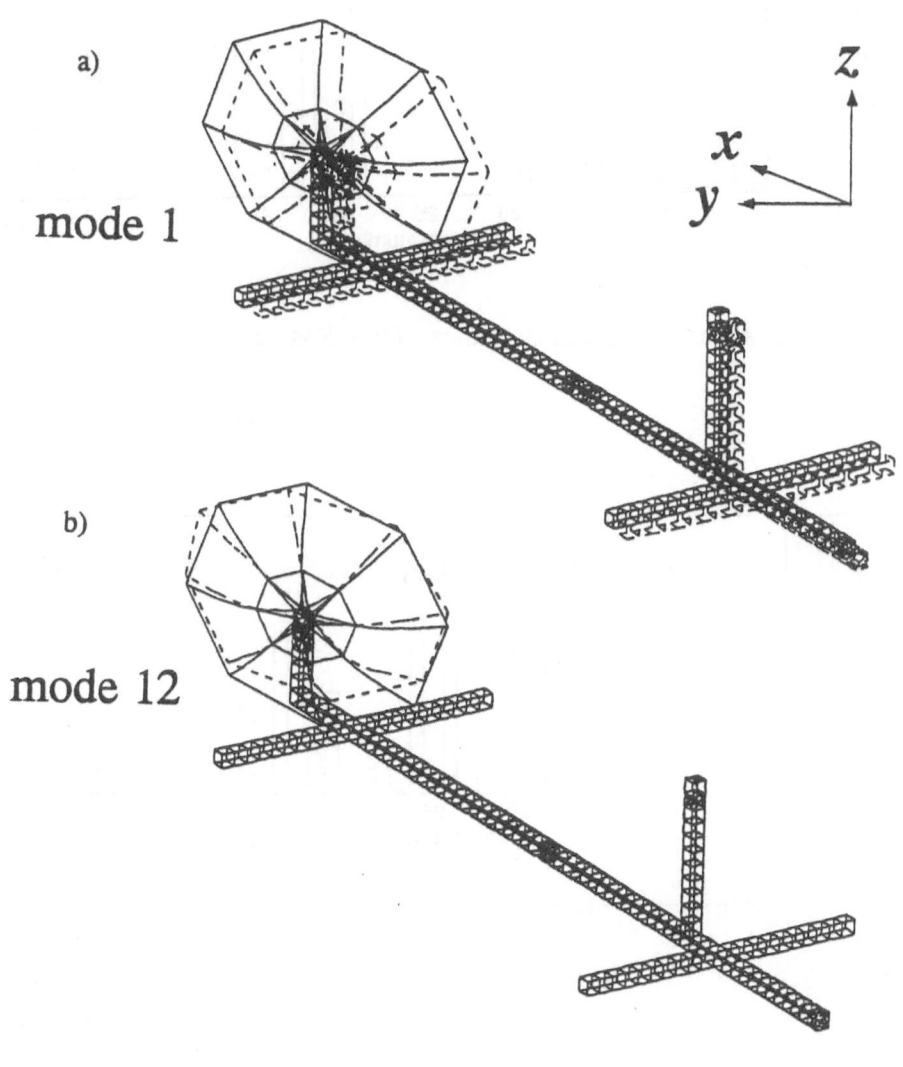

a)

mode 1

b)

mode 12

Figure 5.10: Mode shapes, first (a), and twelfth (b), of CEM

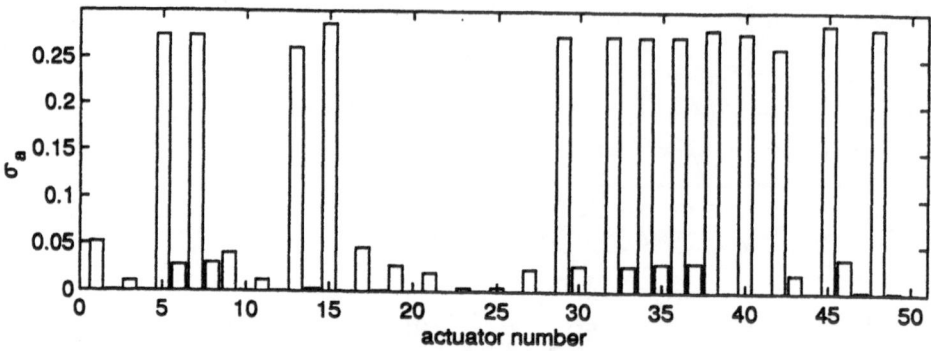

Figure 5.11: Actuator placement indices for CEM, case 2

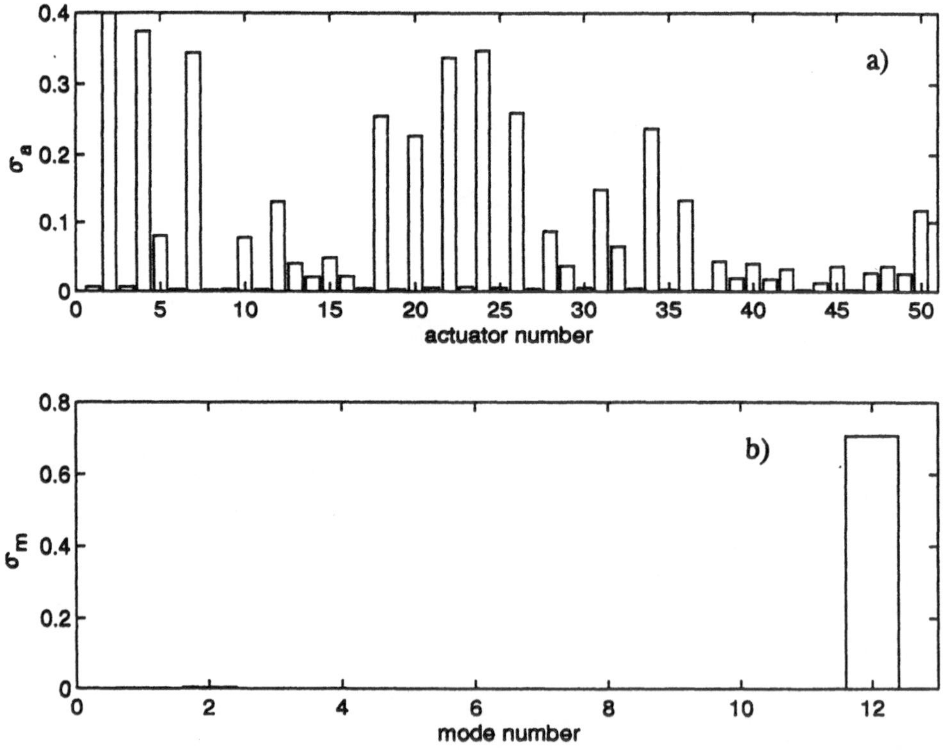

a)

b)

Figure 5.12: CEM indices, case 3 (a) actuator placement, and (b) modal

Chapter 6

Balanced System Identification

The LQG and H_∞ controllers, analyzed later in this book, are model-based ones. The plant model (used as an estimator) is a part of the controller. The performance of the closed-loop system depends, in this case, on the accuracy of the plant model, measured as a discrepancy between the dynamics of the actual plant and its model. For this reason, analytical models of a plant, obtained for example from the finite element model, are acceptable in the simulation stages only. In implementation the test data are used to determine the plant model, in a procedure known as system identification.

System identification is a fairly developed research field, and the reader will find up-to date identification methods in the most comprehensive studies of Natke, Ref.[92], Ljung Ref.[80], Juang Ref.[64], and Ewins, Ref.[27], and get a good insight into the problem. Among many identification procedures available, we describe here only the one which gives the balanced state-space representation. It is in agreement with the main topic of this book, namely balanced system investigations. In addition, the balanced representation gives immediate answer to the question of the order of the identified system. The problem of system order in the identification procedure is an important one: for a system model with too low order a significant part of the plant dynamics is missing; this may cause closed-loop instability due to spillover. A system with too high order, on the other hand, contributes to controller complexity, and may introduce unwanted dynamics and deteriorate the closed-loop system performance.

Balanced system identification is based on the realization method introduced by Ho and Kalman, Ref.[53]. This approach, further developed by Juang and other investigators into the ERA (Eigensystem Realization Algorithm) method, is widely used in flexible structure identification. The ERA method is described in Ref.[64]. The presentation below is based on derivations given in Refs.[44] and [45].

6.1 Discrete Time Balanced Systems

It is most common nowadays that for identification purposes the input and output signals are recorded digitally as discrete time signals. For this reason, we depart in this chapter from the continuous-time models, and we use the discrete-time state-space representation. The sampling time of the recorded signals is denoted by Δt, and the signal u at time $i\Delta t$ is denoted as $u(i)$. With this notation the discrete-time state-space representation is given by the following difference equations

$$x(i+1)=Ax(i)+Bu(i), \qquad y(i)=Cx(i) \tag{6.1}$$

For a system with r inputs and q outputs denote the controllability matrix of order p, $\mathcal{C}_p$ (called also reachability matrix in the discrete-time case), and the observability matrix of order p, O_p,

$$\mathcal{C}_p=[B\ AB\ \dots\ A^{p-1}B], \qquad O_p=\begin{bmatrix} C \\ CA \\ \dots \\ \dots \\ CA^{p-1} \end{bmatrix} \tag{6.2}$$

where $p{\geq}max(r,q)$. These matrices are of dimensions $n{\times}(r{\times}p)$, and $(q{\times}p){\times}n$, respectively, where n is the assumed system order. It is also

138

assumed that $r \times p \geq n$, $q \times p \geq n$. In practice, the sizes $r \times p$ and $q \times p$ of these matrices are much larger than the system order n,

$$r \times p \gg n, \qquad q \times p \gg n \qquad (6.3)$$

in order to minimize the identification error caused by the measurement noise.

The controllability grammian, $W_c(p)$, over the time interval $T=[0 \; p\Delta t]$, and the controllability grammian, $W_o(p)$, over the same interval, are defined as follows

$$W_c(p) = \mathcal{C}_p \mathcal{C}_p^T, \qquad W_o(p) = \mathcal{O}_p^T \mathcal{O}_p \qquad (6.4)$$

In the infinite time interval, i.e., for $t \rightarrow \infty$, and for stable A, the finite time grammians converge to the steady-state grammians, W_c and W_o, which are obtained as the solutions of the following Lyapunov equations

$$A W_c A^T + B B^T = W_c, \qquad A^T W_o A + C^T C = W_o \qquad (6.5)$$

6.2 Identification Algorithms

The presented algorithm is based on the measured impulse responses of a system, and derived from them Markov and Hankel matrices. The matrices h_k of dimension $q \times r$, $k = 0,1,2,\ldots$, such that

$$h_k = C A^k B \qquad (6.6)$$

are called Markov matrices, or Markov parameters. For the discrete-time systems they have a physical interpretation: the ith column of the kth Markov matrix h_k represents the impulse response at the time $k\Delta t$ with a unit impulse at the ith input. Thus, the Markov matrices can be readily measured and therefore are often used in system identification.

The base for the identification algorithm is the Hankel matrix, H_1, and the shifted Hankel matrix, H_2, which are defined as follows

$$H_1 = \begin{bmatrix} h_1 & h_2 & \cdots & h_p \\ h_2 & h_3 & \cdots & h_{p+1} \\ \cdots & \cdots & \cdots & \cdots \\ \cdots & \cdots & \cdots & \cdots \\ h_p & h_{p+1} & \cdots & h_{2p-1} \end{bmatrix}, \quad H_2 = \begin{bmatrix} h_2 & h_3 & \cdots & h_{p+1} \\ h_3 & h_4 & \cdots & h_{p+2} \\ \cdots & \cdots & \cdots & \cdots \\ \cdots & \cdots & \cdots & \cdots \\ h_{p+1} & h_{p+2} & \cdots & h_{2p} \end{bmatrix} \quad (6.7)$$

Their dimensions are $(q \times p) \times (r \times p)$. It is easy to note that the Hankel matrix and the shifted Hankel matrix are obtained from the controllability and the observability matrices, namely

$$H_1 = O_p \mathcal{C}_p, \quad (6.8a)$$

$$H_2 = O_p A \mathcal{C}_p \quad (6.8b)$$

These matrices do not depend on system coordinates. Indeed, let the new representation, x_n, be a linear combination of the representation x, i.e., $x_n = Rx$, then $A_n = RAR^{-1}$, $\mathcal{C}_{np} = R\mathcal{C}_p$, and $O_{np} = O_p R^{-1}$, therefore the Hankel matrices in the new coordinates are the same as in the original ones, since

$$H_{n1} = O_{np} \mathcal{C}_{np} = O_p R^{-1} R \mathcal{C}_p = H_1, \quad (6.9a)$$

$$H_{n2} = O_{np} A_n \mathcal{C}_{np} = O_p R^{-1} R A R^{-1} R \mathcal{C}_p = H_2 \quad (6.9b)$$

In the identification algorithm we do not know, of course, the controllability or observability matrices, $\mathcal{C}_p$ and O_p. However, we have access to the measured impulse responses; consequently, the Hankel matrices, H_1 and H_2, are known. The basic idea in the identification procedure given below is to decompose H_1 similarly to (6.8a), as in Refs.[64], [44], [45]

140

$$H_1 = PQ \tag{6.10}$$

The obtained matrices Q and P serve as the new controllability and observability matrices of the system. Therefore, replacing $\mathscr{C}_p$ and O_p in Eq.(6.8b) with Q and P, respectively, one obtains

$$H_2 = PAQ \tag{6.11}$$

But, if matrices P and Q are full rank, the system matrix A is obtained from the last equation as

$$A = P^+ H_2 Q^+ \tag{6.12}$$

where P^+ and Q^+ are pseudoinverses of P and Q, respectively

$$P^+ = (P^T P)^{-1} P^T, \qquad Q^+ = Q^T (QQ^T)^{-1} \tag{6.13a}$$

such that

$$P^+ P = I, \qquad QQ^+ = I \tag{6.13b}$$

Having determined A, the matrix B of the state-space triple (A,B,C) is easily found as the first r columns of Q (it follows from the definition of the controllability matrix, Eq.(6.2)), therefore

$$B = QE_r, \qquad E_r = [I_r, 0, ...,0]^T \tag{6.14}$$

Similarly, the first q rows of P give the matrix C

$$C = E_p^T P \qquad E_p = [I_p, 0, ...,0]^T \tag{6.15}$$

which completes the identification procedure.

The decomposition (6.10) of the Hankel matrix H_1 is not unique. It

141

could be, for example, the Cholesky, LU, or QR decomposition. However, using the singular value decomposition one obtains the identified triple (A,B,C) in the balanced representation. Indeed, denote the singular value decomposition of the Hankel matrix H_1 as follows

$$H_1 = V\Gamma^2 U^T, \qquad (6.16a)$$

where

$$\Gamma = diag(\gamma_1, \gamma_2, \ldots, \gamma_m) \qquad U^T U = I, \qquad V^T V = I \qquad (6.16b)$$

and $m \le min((q \times p) \times (r \times p))$. By comparing Eqs.(6.10) and (6.16a), one obtains the controllability and observability matrices in the form

$$Q = \Gamma U^T, \qquad P = V\Gamma, \qquad (6.17a)$$

hence their pseudoinverses are, in this case, as follows

$$Q^+ = U\Gamma^{-1}, \qquad P^+ = \Gamma^{-1} V^T \qquad (6.17b)$$

But, from Eq.(6.4) it follows that

$$W_c(p) = QQ^T = \Gamma^2, \qquad W_o(p) = P^T P = \Gamma^2 \qquad (6.18)$$

i.e., that the controllability and observability grammians over the time interval $T = [0 \ p\Delta t]$ are equal and diagonal, hence the system is balanced over the interval T. This fact has a practical meaning: the states of the identified system are equally controlled and observed. But, weakly observed *and* weakly controlled states can be ignored, since they do not contribute significantly to the system dynamics. They are usually below the level of measurement noise. Thus, using the singular value decomposition of the Hankel matrix H_1, one can readily determine the order of the identified triple (A,B,C): the states with small singular values can be truncated. Of course, the measurement and system noises have an impact on the Hankel singular values, and this problem is analyzed in Ref.[64].

Besides the noise impact on the identification accuracy, one has to carefully determine the sampling time Δt: it should be small enough to include the system bandwidth, but not too small, in order to ease the computational burden. The size of the record p should satisfy the conditions of Eq.(6.3). The procedure identifies the model such that the model response fits the plant response for the time $T=[0\ p\Delta t]$. Thus too short records can produce a model that response fits to the plant response within the time segment T, and departs outside T, which makes the model unstable. This will be illustrated in the following examples.

6.3 Applications

The identification algorithm is checked for the simple structure, the truss, and the Deep Space Network antenna.

Simple structure (Fig.2.1). The parameters are $m_1=m_2=m_3=1$, $k_1=k_2=k_3=k_4=6$, $d_1=d_2=d_3=d_4=0.024$; the single input is applied to the mass m_3, and the single output is the velocity of the mass m_2. Since the analytical model is used to simulate data, the system order $n=6$ is known in advance. The continuous time system representation is

$$A_o = \begin{bmatrix} 0 & 0 & 0 & 1 & 0 & 0 \\ 0 & 0 & 0 & 0 & 1 & 0 \\ 0 & 0 & 0 & 0 & 0 & 1 \\ -12 & 6 & 0 & -0.048 & 0.024 & 0 \\ 6 & -12 & 6 & 0.024 & -0.048 & 0.024 \\ 0 & 6 & -6 & 0 & 0.024 & -0.024 \end{bmatrix}, \quad B_o = \begin{bmatrix} 0 \\ 0 \\ 0 \\ 0 \\ 0 \\ 1 \end{bmatrix},$$

$$C_o = [0\ 0\ 0\ 0\ 1\ 0].$$

For this single-input, single-output case, i.e., for $q=r=1$, we assume the Hankel matrix order $p=20$. The sampling time is $\Delta t=0.05$ sec, and the

143

first 20 impulse responses (Markov parameters) are given in Table 6.1.

Table 6.1. Markov parameters (MP) for the simple system.

sample	1	2	3	4	5	6	7	8	9	10
MP	.0002	.0010	.0025	.0046	.0072	.0103	.0137	.0171	.0206	.0238

Table 6.2. Hankel singular values (HSV) of the simple system.

state	1	2	3	4	5	6	7	8	9	10
HSV	.0059	.0048	.0019	.0007	.00006	.00001	$\sim 10^{-9}$	$\sim 10^{-9}$	$\sim 10^{-9}$	$\sim 10^{-9}$

For these parameters the Hankel matrix H_1, as in Eq.(6.7), was generated, and its singular values γ_i, $i=1,2,...,10$ are given in Table 6.2. The table shows the rapid drop in γ_i for $i \geq 7$, noting that the system order is $n=6$. The identified discrete-time system triple (A,B,C) is obtained from Eqs.(6.12), (6.14) and (6.15)

$$
A = \begin{bmatrix}
1.0601 & -0.1511 & 0.0272 & -0.0046 & 0.0006 & -0.0001 \\
0.1511 & 1.0473 & -0.1875 & 0.0099 & -0.0019 & 0.0002 \\
0.0272 & 0.1875 & 0.8361 & -0.0649 & 0.0039 & -0.0008 \\
0.0046 & 0.0099 & 0.0649 & 1.1451 & -0.1624 & 0.0135 \\
0.0006 & 0.0019 & 0.0039 & 0.1624 & 0.9936 & -0.1599 \\
0.0001 & 0.0002 & 0.0008 & 0.0135 & 0.1599 & 0.8271
\end{bmatrix}, \quad
B = \begin{bmatrix}
0.0767 \\
-0.0836 \\
0.0354 \\
-0.0013 \\
0.0001 \\
-0.00005
\end{bmatrix},
$$

$$
C = [0.0767 \quad 0.0836 \quad 0.0354 \quad 0.0013 \quad 0.0001 \quad 0.00005]
$$

This representation is balanced.

Truss (Fig.2.2). The input is a vertical force applied at node $n4$; the output is the rate collocated with the input force. The sampling time $\Delta t=0.02$ sec was chosen for the signal recording. The system identification was conducted for different record lengths, namely $t_{30}=30\Delta t=0.6$ sec, $t_{50}=50\Delta t=1$ sec, $t_{100}=100\Delta t=2$ sec, and $t_{200}=200\Delta t=4$

sec,. The results are shown in Figs.6.1a,b. Fig.6.1a shows how the identified system order depends on the length of the record. Although from the algorithm it follows that the system can be identified for records longer than the system order, the limited-time grammians show that the identified model response will match the plant response for the time of the record length. This is shown in Fig.6.1b, where the the system for the time $t_{30}=30\Delta t=0.6$ sec matches the plant response up to 2 sec, and for the record of $t_{50}=50\Delta t=1$ sec duration it matches up to 3 sec. The records $t_{100}=2$ sec, and $t_{200}=4$ sec, are long enough to also include the slow components, thus matching the entire time period.

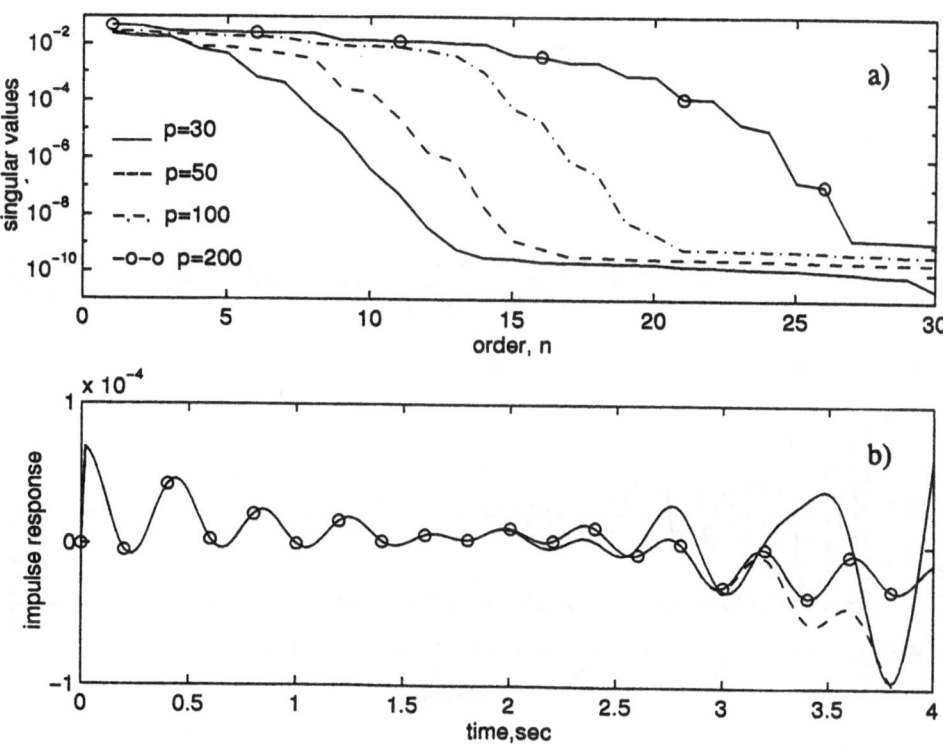

Figure 6.1: Truss Hankel singular values (a), and impulse responses (b).

The DSN antenna (Figs.2.3, 2.4). The antenna has two outputs: elevation and azimuth angles, and a single input: elevation rate. The sampling time was $\Delta t = 0.02$ sec. The record length $p=300$ was chosen. The Hankel singular values for the identified model are shown in Fig.6.2a, indicating the model order as $n=31$ (actually, it was 35, but some states evidently were not excited and observed). The plant impulse responses are shown in solid lines in Fig.6.2b, and the impulse responses of the identified model in dashed lines, showing good coincidence.

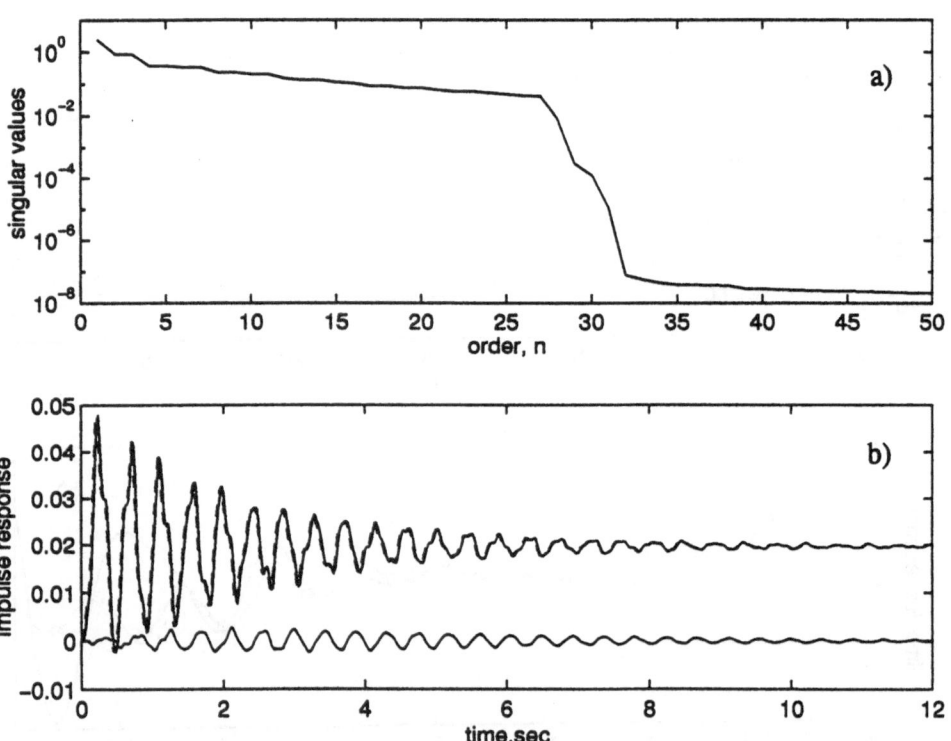

Figure 6.2: Hankel singular values (a), and impulse responses for the DSN antenna (b).

Chapter 7

Balanced Dissipative Controllers

The most straightforward approach to the design of controllers for flexible structures is to implement a direct proportional gain between the input and the output. This approach seldom gives superior performance in a general case, since the performance enhancement often causes the reduction of the stability margin. However, if some conditions are satisfied, one obtains a special type of proportional controllers - dissipative ones. The dissipative controllers are often implemented since, as mentioned by Joshi [62, p.45] "the stability of dissipative controllers is guaranteed regardless of the number of modes controlled (or even modeled), and regardless of parameter errors." However, the simplicity of the control law does not simplify the design. Typically, in order to obtain required performance typically a large set of controllers have to be implemented. Determining the gains for this set is not an obvious task. In this chapter we will investigate the properties of the dissipative controllers, and will show the design of dissipative controllers for flexible structures to meet certain objectives.

7.1 Definition and Properties

The dissipative controllers and their properties are based on the Popov's theory of hyperstability, Ref.[95], which was subsequently developed as a positive real property of the control systems Refs.[2],

[9], and as the dissipative (passive) property of the systems Refs.[113], [114], and [22]. The terms: dissipative, passive, positive real, and hyperstable systems are synonyms in our applications, and their inter-relations are discussed by Wen, Ref.[111]. In this chapter the above systems will be called dissipative systems.

Consider a square stable plant (A,B,C), i.e., a linear system with the number of inputs equal to the number of outputs. An open-loop square system with simple poles is dissipative, see Ref.[2], if there exist a symmetric positive definite matrix P and a matrix Q that satisfy the following equation

$$A^{\mathrm{T}}P+PA=-Q^{\mathrm{T}}Q, \qquad B^{\mathrm{T}}P=C \tag{7.1}$$

The system is strictly dissipative if $Q^{\mathrm{T}}Q$ is positive definite.

The definition (7.1) allows for simple determination of a dissipative system. For a given A and B, the matrix Q is selected. Next, one solves the first of Eq.(7.1) for P, and the output matrix C is found from the second equation (7.1).

Three particular cases of the dissipative systems are discussed. In the first case $Q=B^{\mathrm{T}}$ is chosen to obtain

$$P=W_{\mathrm{c}}, \quad \text{and} \quad C=B^{\mathrm{T}}W_{\mathrm{c}}^{-1} \tag{7.2a}$$

where W_{c} is the controllability grammian. In this case the sensors are collocated with actuators and the actuators are weighted inversely to the system controllability grammian. In the second case the matrix $Q=CW_{\mathrm{o}}^{-1}$ is chosen, where W_{o} is the observability grammian. Thus

$$P=W_{\mathrm{o}}^{-1}, \quad \text{and} \quad B=W_{\mathrm{o}}C^{\mathrm{T}} \tag{7.2b}$$

In this case the actuators are collocated with sensors, and the sensors are weighted proportionally to the system observability grammian. If a system is balanced, then in cases 1 and 2 one obtains $C=B^T\Gamma^{-2}$, where Γ is a diagonal matrix of the Hankel singular values.

In the third case $Q=(-A-A^T)^{1/2}$ is chosen, therefore one obtains

$$P=I, \quad \text{and} \quad B=C^T \qquad (7.2c)$$

In this particular case the actuator and sensors are collocated. This case is most frequently used, since it requires simple actuator and sensor collocation without weighting.

The guaranteed stability of the closed-loop system is the most useful property of the dissipative system. It was shown by Desoer, Ref.[22], and by Benhabib, Iwens and Jackson, Ref.[9] that, for the square and strictly dissipative plant and the square and dissipative controller (or *vice versa*: the square and dissipative plant and the square and strictly dissipative controller), the closed-loop system is asymptotically stable. In particular, if the feedback gain matrix is a positive definite, the closed-loop system is asymptotically stable.

As a corollary, consider a dissipative system with the state-space representation (A,B,C) which has collocated sensors and actuators, that is, $C=B^T$. In this case a closed-loop system with the feedback gain

$$u=-Ky, \qquad (7.3)$$

where $K=diag(k_i)$, and $k_i>0$, is stable. This particularly useful configuration can be used only if there is the freedom to choose the collocated sensors and actuators, and if the number of the available sensors and actuators is large enough to satisfy the performance requirements.

The stability property allows one to design simple and stable controllers, regardless of the plant parameter variations. However, one has to be aware, that stability does not imply performance, and sometimes the performance of these controllers can be poor, as reported by Hyland in Ref.[56].

7.2 Balanced Dissipative Controller for Flexible Structures

Consider a flexible structure (A,B,C) with an equal number of sensors and actuators. According to the previously introduced property, the collocation of sensors and actuator and positive definite gain makes it dissipative. However, flexible structures inherit a property which restricts the collocation of sensors and actuators. The restriction follows from the state-space representation of a flexible structure, as in Eq.(2.4), or in Eq.(2.11). In these representations the upper half of the matrix B is equal to zero, and this zero part is a generic one, that is, it is independent of the applied forces (it would be non-zero only if a force impulse $f\Delta t$ would be applied to the input). Thus, in order to satisfy the collocation requirement, the left half of C must be equal to zero. But the displacement measurements are located in this part (the right part locates the displacement rates). Consequently, a flexible structure is dissipative if the force inputs and the rate outputs are collocated.

When designing the dissipative controllers, one has to collocate the sensor and actuator. When working with an analytical model it is beneficial to chose the inputs first, and determine the outputs subsequently by introducing $C=B^T$. In this way a physically realizable dissipative system is obtained. In the case when the outputs are determined first and the inputs are picked afterwards as $B=C^T$, one still deals with a dissipative system, but not necessarily a physically realizable one. Note also that the collocation of force actuator and rate sensors is a sufficient but by no means necessary condition of dissipativeness. For example, if the weighted collocation is used, as

150

in Eqs.(7.2a,b), the system with displacement sensors is dissipative.

In order to determine the properties of the dissipative balanced structures consider further the dissipativity conditions (7.1) for a flexible structure in the balanced coordinates in the modal form 2. Consider also a feedback as in Eq.(7.3), and let the number of inputs and outputs be p. In this case the closed-loop equations are as follows

$$\dot{x} = (A-BKC)x + BKu_o, \quad y = Cx \qquad (7.4)$$

where u_o is a control command ($u_o=0$ in the case of vibration suppression). Since the matrix A is in the modal form 2, and K is diagonal, $K=diag(k_1,...k_p)$, then in the balanced coordinates with collocated sensors and actuators one obtains the closed-loop matrix $A_c=A-BKB^T$ in the form

$$A_c = A - \sum_{j=1}^{p} k_j B_j B_j^T \qquad (7.5)$$

where B_j is the j-th column of B. In the balanced coordinates matrix A_c is diagonally dominant, that is, $A_c \cong diag(A_{c1},...A_{cn})$, where A_{ci} is 2×2 block. For this block the Eq.(7.5) is as follows

$$A_{ci} \cong A_i - \sum_{i=1}^{p} k_j B_{ji} B_{ji}^T \qquad (7.6a)$$

In this equation the cross terms $B_{jk}B_{ji}^T$ (for $k \neq i$) were omitted as negligible in the balanced coordinates, see Eq.(3.61d), where B_{ji} is the i-th block of the j-th column of B. Also, for the balanced system the following holds $B_{ji}B_{ji}^T \cong -\gamma_{ji}^2(A_i+A_i^T)$, see Eq.(3.61d), where γ_{ji} is the i-th Hankel singular value obtained for the j-th column of B, i.e., for the triplet (A,B_j,B_j^T). Thus, Eq.(7.5) is now

$$A_{ci} \cong A_i + 2 \sum_{j=1}^{p} k_j \gamma_{ji}^2 (A_i + A_i^T), \tag{7.6b}$$

and introducing A_i as in Eq.(2.13b), and noting that $A_i + A_i^T = -2\zeta_i \omega_i I_2$, as in Eq.(3.61d), one re-writes Eq.(7.6)

$$A_{ci} \cong \begin{bmatrix} -\beta_i \zeta_i \omega_i & -\omega_i \\ \omega_i & -\beta_i \zeta_i \omega_i \end{bmatrix}, \tag{7.7a}$$

with the parameter β_i given as

$$\beta_i = 1 + 2 \sum_{j=1}^{p} k_j \gamma_{ji}^2 \tag{7.7b}$$

Comparing the closed-loop matrix as in Eq.(7.7a) and the open-loop matrix as in Eq.(2.13b) it can be seen that β_i is a measure of the shift of the i-th pair of poles. Denote the closed-loop pair of poles $(\lambda_{cri}, \pm j\lambda_{cii})$ and the open-loop pair $(\lambda_{ori}, \pm j\lambda_{oii})$, then it follows from Eq.(7.7) that they are related

$$(\lambda_{cri}, \pm j\lambda_{cii}) \cong (\beta_i \lambda_{ori}, \pm j\lambda_{oii}), \qquad i = 1, \ldots, n \tag{7.8}$$

where β_i is defined in Eq.(7.7b). Eq.(7.8) shows that the real part of the i-th pair of poles is shifted, while the imaginary part is stationary. The shift is proportional to the gain of each input, and to the i-th Hankel singular values associated with each input.

Eq.(7.7) sets the basic limitation for the dissipative controller design: that the number of inputs (and outputs) limits the number of controlled modes (or controlled pairs of poles). In order to illustrate it, assume a single-input-single-output system. In this case $\beta_{1i} = 1 + 2k_1 \gamma_{1i}^2$, and the scalar gain k_1 is the only free parameter available for the design. Thus only one pole can be shifted to the required position. If more than one pair should be shifted, their placement would be a least-square compromise, which typically would be non-satisfactory. Thus, in order to avoid this rough approximation, it is often required for the dissipative controllers to have a large number of sensors and actuators to meet the required performance criteria.

The pole-shift coefficient β_i can be also interpreted as a ratio of the open- and closed-loop Hankel singular values, or as a ratio of the variances of open-loop (σ_{oi}^2) and closed-loop (σ_{ci}^2) states excited by the white noise input, i.e.,

$$\beta_i \cong \frac{\gamma_{oi}^2}{\gamma_{ci}^2} = \frac{\sigma_{oi}^2}{\sigma_{ci}^2} \tag{7.9}$$

Since $\beta_i \geq 1$, therefore it is a relative measure of the noise suppression of the closed-loop system with respect to the open-loop system. This interpretation follows from the closed-loop Lyapunov equation

$$(A-BKB^T)\Gamma_c^2 + \Gamma_c^2(A-BKB^T)^T + BB^T = 0 \tag{7.10a}$$

which for the i-th pair of variables is as follows

$$(A_i - \sum_{j=1}^{p} k_i B_{ji} B_{ji}^T)\gamma_{ci}^2 + \gamma_{ci}^2(A_i - \sum_{j=1}^{p} k_i B_{ji} B_{ji}^T)^T + B_i B_i^T \cong 0 \tag{7.10b}$$

Introducing Eq.(3.61d), after some algebra, one obtains

$$\gamma_{ci}^2 + 2\gamma_{ci}^2(\sum_{j=1}^{p} k_j \gamma_{ji}^2) - \gamma_{oi}^2 \cong 0 \tag{7.11}$$

or, finally

153

$$\frac{\gamma_{oi}^2}{\gamma_{ci}^2} \cong 1 + 2 \sum_{j=1}^{p} k_j \gamma_{ji}^2 = \beta_i \qquad (7.12)$$

Based on Eqs.(7.7b), (7.8), and (7.9) a tool for pole placement of the dissipative controllers is developed. The task is to determine gains k_j, $j=1,...,p$, such that the selected poles are placed at the required location (or as close as possible in the least-square sense). Equivalently, the task is to determine gains k_j, $j=1,...,p$, such that the input noise of the selected modes is suppressed at ratio β_i. The approach follows from Eq.(7.7b), since one can determine the gains such that q poles are shifted by β_i, $i=1,...,q$, i.e., $\lambda_{cri}=\beta_i\lambda_{ori}$, or the noise be suppressed by β_i, i.e., $\sigma_{oi}^2=\beta_i\sigma_{ci}^2$. Define the gain vector k

$$k=[k_1, \ k_2,...,k_p]^T \qquad (7.13)$$

so that Eq.(7.7b) can be re-written as

$$d\beta \cong Gk \qquad (7.14)$$

where $d\beta$ is the vector of pole shifts

$$d\beta = \begin{bmatrix} \beta_1 - 1 \\ \beta_2 - 1 \\ .... \\ .... \\ \beta_q - 1 \end{bmatrix} \qquad (7.15a)$$

and G is the matrix of the system Hankel singular values for each actuator and sensor location

$$G = 2 \ [\gamma_1 \ \gamma_2 \ ... \ \gamma_p] = 2 \begin{bmatrix} \gamma_{11}^2 & \gamma_{21}^2 & \cdots & \gamma_{p1}^2 \\ \gamma_{12}^2 & \gamma_{22}^2 & \cdots & \gamma_{p2}^2 \\ \cdots & \cdots & \cdots & \cdots \\ \gamma_{1q}^2 & \gamma_{2q}^2 & \cdots & \gamma_{pq}^2 \end{bmatrix} \qquad (7.15b)$$

154

where $\gamma_i = [\gamma_1, \gamma_2, ..., \gamma_q]^T$ is the set of the Hankel singular values for the i-th actuator/sensor location, and γ_{ij} is the j-th Hankel singular value for the i-th actuator/sensor location.

The least-square solution of Eq.(7.14) is obtained

$$k \cong G^+ d\beta \qquad (7.16)$$

where G^+ is pseudoinverse of G. The set of equations (7.14) is either overdetermined $(q>p,$ or $rank(G)=p,)$, or square $(q=p=rank(G))$, or underdetermined $(q<p,$ or $rank(G)=q)$, see Ref.[50]. The form of pseudoinverse depends on the number of inputs and outputs p, and the number of poles shifted, q, that is, on the rank of the matrix G.

7.3 Design Examples

Two examples of balanced dissipative controller design are presented: the controller design for the simple flexible system, and for the truss structure. The dissipative controller for the Deep Space Network antenna is not investigated, since the antenna pre-defined inputs and outputs are not suitable for this kind of design.

7.3.1 Simple Structure

The system is shown in Fig.2.1, with the masses $m_1=m_2=m_3=1$, stiffness $k_1=10, k_2=k_4=3, k_3=4$, and the damping matrix D as a linear combination of the mass and stiffness matrices, $D=0.004K+0.001M$. The input force is applied to the mass m_3, the output is the rate of the same mass. The poles of the open-loop system are

$\lambda_{o1,o2}=-0.0024\pm j0.9851,$

$\lambda_{o3,o4}=-0.0175\pm j2.9197,$

$\lambda_{o5,o6}=-0.0295\pm j3.8084.$

The system open-loop balanced representation was obtained with the following vector of Hankel singular values

$$\gamma_i^2 = [63.6418,\ 63.6413,\ 4.9892,\ 4.9891,\ 0.2395,\ 0.2391]^\mathrm{T}$$

It is required to shift the first pole by increasing its damping twofold, and leave the other poles stationary. For this shift coefficients are $\beta_1 = 2$, and $\beta_2 = \beta_3 = 1$, therefore $d\beta = [1\ 1\ 0\ 0\ 0\ 0]^\mathrm{T}$. Also for this case $G = 2\gamma_i^2$, thus the gain $k = 0.0078$ is obtained from Eq.(7.16). For this gain, the closed-loop eigenvalues were computed

$$\lambda_{c1,c2} = -0.0049 \pm j0.9851,$$
$$\lambda_{c3,c4} = -0.0189 \pm j2.9197,$$
$$\lambda_{c5,c6} = -0.0296 \pm j3.8084$$

and from this result one can see that the actual shifts $\beta_1 = 1.9939$, $\beta_2 = 1.0779$, and $\beta_3 = 1.0037$ are close to the required ones.

Next, consider a design which increases the first and the second pole damping twofold, and leaves the third stationary. In this case $\beta_1 = \beta_2 = 2$, and $\beta_3 = 1$ is required, therefore $d\beta = [1\ 1\ 1\ 1\ 0\ 0]^\mathrm{T}$. The gain $k = 0.0084$ is obtained from Eq.(7.16), and the closed-loop eigenvalues for this gain are computed:

$$\lambda_{c1,c2} = -0.0051 \pm j0.9851,$$
$$\lambda_{c3,c4} = -0.0190 \pm j2.9197,$$
$$\lambda_{c5,c6} = -0.0296 \pm j3.8084.$$

Comparing the open- and closed-loop poles, one can see that the actual shifts, $\beta_1 = 2.0718$, $\beta_2 = 1.0840$, and $\beta_3 = 1.0040$, are almost the same as in the first case (small shift in the second pole is observed). Thus, we hardly met the requirements. This case shows that for the underdetermined problem (number of inputs is smaller than the number of poles to be shifted), the obtained least square result is not

156

satisfactory, although it is the best one can get in the given situation.

7.3.2 Truss

The truss is presented in Fig.2.2. A single control force is applied at node $n7$, directed vertically. The output is a rate collocated with the force. The system has 26 balanced states. The most controllable and observable mode is suppressed by increasing the damping of this mode two times. The required feedback gain, as in Eq.(7.3), is obtained from Eq.(7.16). In order to do this, note that in this case $\beta_1=2$, and the remaining β's are equal to 1. Let γ_1 be a vector of the Hankel singular values for the system, then $G=\gamma_1$. For this case $d\beta=[2\ 2\ 1\ 1\ 0\ 0\ ...0]$, and one obtains the gain $k=1548.5$ from Eq.(7.16). For this gain the closed-loop poles were determined, and the pole shift was obtained as a ratio of real parts of closed- and open-loop poles, as in definition Eq.(7.8), i.e. $\beta_i=\lambda_{cri}/\lambda_{ori}$. The plot of β_i is shown in Fig.7.1. It shows that $\beta_1=1.9978\cong2$, as required. The damping of the first pole increased two times, while the other poles changed insignificantly.

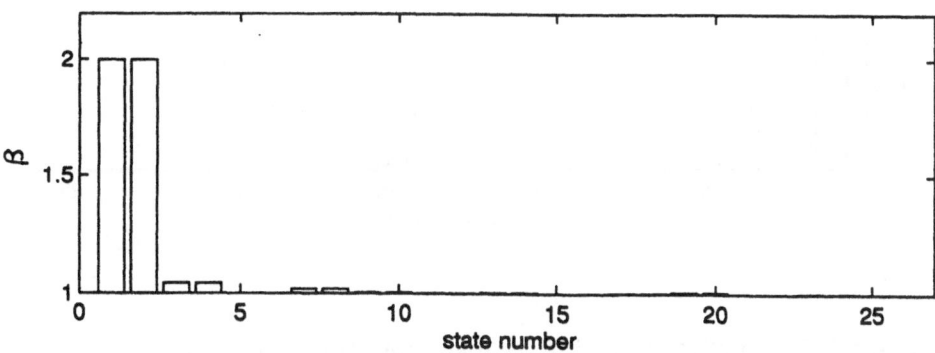

Figure 7.1: Coefficient β_i for the single-input-single-output truss

Next, two control forces are applied at node *n7*. The first one is directed vertically, the second one, horizontally. The outputs are the collocated rates. The disturbance force *d* at node *n8*, at *y*-direction, was added. The suppression of vibrations due to disturbances is required. It is done by introducing pole shift of 300 for the first mode, and of 100 for the second mode, obtaining

$$d\beta = [299, \ 299, \ 99, \ 99, \ 0 \ , ... 0].$$

For this case, one obtains the gains $k = [3.127, \ 2.688] \times 10^5$ from Eq.(7.16). The shift for the first mode was 307, and for the second mode 111 (see Fig.7.2), i.e., close to the required one. The open- and the closed-loop system response to the white noise disturbance *d* is shown in Fig.7.3, where the improved disturbance rejection property is observed.

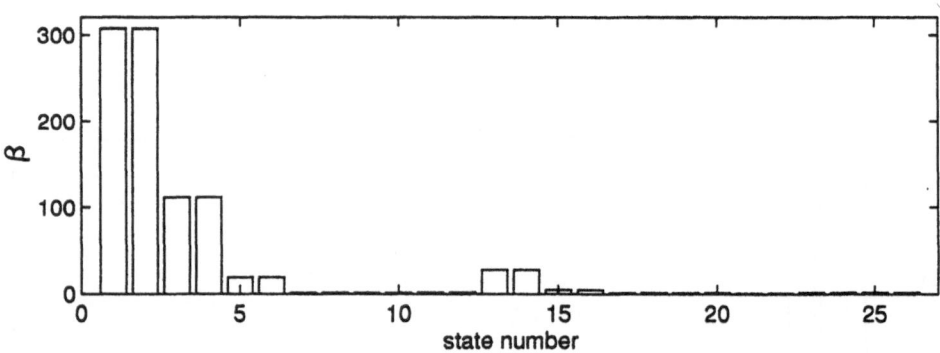

Figure 7.2: Coefficient β_i for the two-input-two-output truss

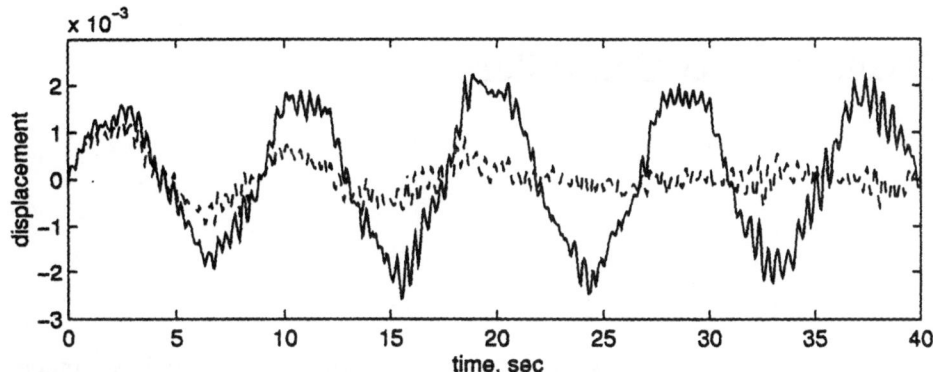

Figure 7.3: Open- and closed-loop output due to disturbances

159

Chapter 8

Balanced LQG Controllers

Recently the control issues for flexible structures have gained much attention, particularly in the space applications. The growing interest reflects the efforts to maintain high precision positioning of ever lighter and more flexible structures. The linear quadratic gaussian (LQG) controllers are often implemented for tracking and disturbance rejection purposes. There have been many investigations into the analysis and design of the LQG controllers, and a good insight into the variety of approaches can be obtained from Kwakernaak and Sivan [70], Maciejowski [84], Anderson and Moore [3], Furuta and Sano [30], and Lin [78]. In this chapter a balanced LQG controller for flexible structures is described.

The two essential issues in the LQG controller design are emphasized: the determination of the weights of the performance index, and the controller order reduction. Despite the already extensive research, both problems require further investigations. They are especially important for the flexible structures, since the structure models are typically of high order, making the weight determination a difficult task, and order reduction a necessity.

The performance of the LQG controller depends on the choice of the weights. If the weights of the LQG performance index are inappropriately chosen, the LQG controller performance will not reflect the requirements. Although the selection of weights is a crucial issue, it is, most often, not an easy task. As stated by Lin [78, p.93] "It takes a great deal of experience to transform design requirements and objectives to the performance index that will produce the desired performance." Our task is to replace experience with analytical tools.

On the other hand, order reduction of the controller is equally important in implementation. The order of the controller is equal to the order of the plant, which is in most cases unacceptably high. Typically for flexible structures it ranges from 200 to 2000. In implementation it should be low, usually less than 30.

In this book both problems, weight determination, and the controller order reduction, are solved in the LQG balanced coordinates. The LQG balanced representation differs from the open-loop balanced representation discussed earlier. The gains of an LQG controller are obtained from the solutions of the controller Riccati equation (CARE) and the estimator Riccati equation (FARE). Similarly to the open-loop balanced representation, where the solutions of the two Lyapunov equations are equal and diagonal, the solutions of CARE and FARE equations in the LQG balanced representation are equal and diagonal. Their diagonal entries are the LQG characteristic values of the system, as defined in Ref.[60]. The LQG balanced representation and the properties of the flexible structures is a fruitful combination which allows to solve efficiently the two mentioned problems.

8.1 Definition and Properties

In the LQG controller design we consider a situation where a stable plant is perturbed by random disturbances, and only some of the states are measured through the noise-corrupted output. A block diagram of an LQG controller is shown in Fig.8.1, where the plant is described as follows

$$\dot{x} = Ax + Bu + v, \qquad y = Cx + w \qquad (8.1)$$

The process noise, v, has intensity $V=E(vv^T)$, the measurement noise, w, has intensity $W=E(ww^T)$, and the noises v and w are uncorrelated, $E(vw^T)=0$, where $E(.)$ is an expectation operator. It is assumed that $W=I$ without loss of generality. The plant states x are approximated with estimator states $\hat{x}$. The performance of the system is determined with the estimator gain (K_e) and the controller gain (K_p). Thus the task is to determine the controller and estimator gains such that the performance index J

$$J^2 = E\left[\int_0^\infty (x^T Qx + u^T Ru)dt\right] \qquad (8.2)$$

is minimal, where R is a positive definite input weight matrix, and Q a positive semi-definite state weight matrix. It is assumed $R=I$ without loss of generality.

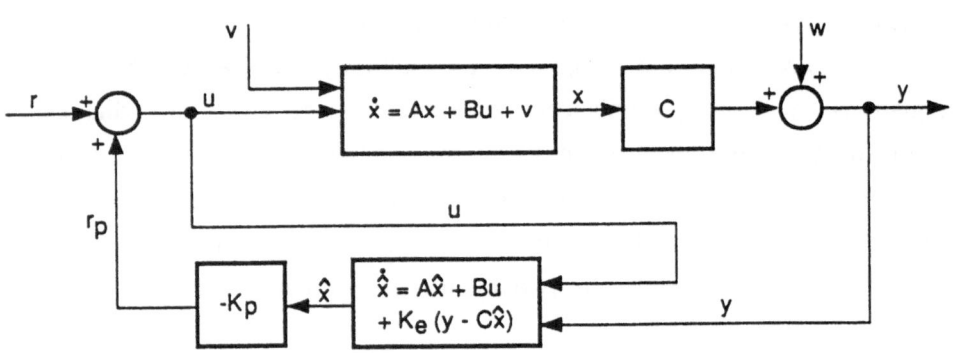

Figure 8.1: LQG controller

The minimum of J is obtained for the feedback $u=-K_p\hat{x}$, with the gain matrix, $K_p=B^T S_c$, and S_c is the solution of the controller Riccati equation (CARE), see for example Ref.[70]

$$A^T S_c + S_c A - S_c BB^T S_c + Q = 0 \qquad (8.3a)$$

The optimal estimator gain is given by $K_e = S_e C^T$, where S_e is the solution of the estimator Riccati equation (FARE)

$$AS_e + S_e A^T - S_e C^T C S_e + V = 0. \tag{8.3b}$$

Jonckheere and Silverman [60], and Opdenacker and Jonckheere [93] have shown for the case of $Q = C^T C$, and $V = BB^T$, and Gawronski [34] for a general case that there exists a diagonal positive definite $M = diag(\mu_i)$, $i = 1, ..., n$, $\mu_i > 0$, such that

$$S_c = S_e = M = diag(\mu_1, \mu_2, ..., \mu_n), \qquad \mu_1 \geq \mu_2 \geq ... \geq \mu_n > 0 \tag{8.4}$$

A state-space representation with the condition (8.4) satisfied is an LQG balanced representation, and μ_i, $i = 1, ..., n$ are its LQG characteristic values.

Let R be the transformation of the state x such that $x = R\bar{x}$. Then the solutions of CARE and FARE, and the weighting matrices in the new coordinates, are as follows

$$\bar{S}_c = R^T S_c R, \quad \bar{S}_e = R^{-1} S_e R^{-T}, \tag{8.5a}$$

$$\bar{Q}_c = R^T Q_c R, \quad \bar{Q}_e = R^{-1} Q_e R^{-T} \tag{8.5b}$$

The transformation R to the LQG balanced representation is obtained as follows. Decompose S_c and S_e

$$S_c = P_c^T P_c, \quad S_e = P_e P_e^T \tag{8.6}$$

and form a matrix H

$$H = P_c P_e \tag{8.7}$$

Find the singular value decomposition of H

$$H = V M U^T \qquad (8.8)$$

then the transformation matrix is obtained as

$$R = P_e U M^{-1/2} \quad \text{or} \quad R = P_c^{-1} V M^{1/2} \qquad (8.9)$$

Introduction of R to Eq.(8.5a) shows that Eq.(8.4) is satisfied.

The Matlab function *bal_LQG* which transforms a representation (A,B,C) to the LQG balanced representation (A_b, B_b, C_b) is given in the Appendix B4.

8.2 Weights of Special Interest

Here, weighting matrices of special form, and the corresponding balanced solutions will be considered. First, for a fully controllable system, consider the weights Q and V as follows

$$Q = 2W_c^{-1} B B^T W_c^{-1}, \quad V = 2W_o^{-1} C^T C W_o^{-1} \qquad (8.10)$$

In this case one obtains the inverses of the controllability and the observability grammians as the CARE and FARE solutions

$$S_c = W_c^{-1}, \quad S_e = W_o^{-1} \qquad (8.11)$$

It can be proved by introduction of Eqs.(8.10) and (8.11) into CARE, which gives

$$A^T S_c + S_c A + S_c B B^T S_c = 0 \qquad (8.12)$$

for $S_c = W_c^{-1}$, which is a Lyapunov equation (3.3). Similar proof can be shown for the solution of FARE.

The weights as in Eq.(8.10) penalize each state reciprocally to its degree of controllability and observability. Particularly, when the weights Q, and V are determined in the open-loop balanced representation, one obtains from Eq.(8.10) that the system is LQG balanced, with $S_c = S_e = M = \Gamma^{-2}$.

Next, define the closed-loop performance. Let $\Gamma_o = diag(\gamma_{oi})$, and $\Gamma_c = diag(\gamma_{ci})$, $i = 1,...,n$, be the matrices of open- and closed-loop Hankel singular values. The matrix B

$$B = \Gamma_o^2 \Gamma_c^{-2} = diag(\beta_i) = diag(\gamma_{oi}^2 / \gamma_{ci}^2) \qquad (8.13)$$

is the ratio of open-loop (γ_{oi}) and closed-loop (γ_{ci}) Hankel singular values, or the ratio of open- and closed-loop state variances excited by the white noise input. Thus B is a measure of the closed-loop performance.

For the weights as in Eq.(8.10), and a system in the open-loop balanced coordinates, one obtains each state of the plant and the estimator equally influenced by the feedback. As a result

$$B = 3I \qquad (8.14)$$

This follows from the Lyapunov equation for the closed-loop controllability grammian W_c, which is obtained from the following equation

$$(A - BB^T S_c)W_c + W_c(A^T - S_c BB^T) + BB^T = 0 \qquad (8.15a)$$

According to Eq.(8.11), $S_c = \Gamma_o^{-2}$, and introducing $W_c = \Gamma_o^2/3$ to Eq.(8.15a), one obtains

$$A\Gamma_o^2 + \Gamma_o^2 A^T + BB^T = 0 \qquad (8.15b)$$

which shows that $W_c = \Gamma_o^2/3$ is a solution of Eq.(8.15a), and consequently that $B = 3I$.

Consider another set of weights of a fully controllable system, namely

$$Q = C^T C + W_o B B^T W_o, \qquad V = B B^T + W_c C^T C W_c \qquad (8.16)$$

then one obtains the observability and controllability grammians as solutions of CARE and FARE

$$S_c = W_o, \qquad S_e = W_c \qquad (8.17)$$

This can be proved simply by introduction of Eq.(8.16) to the CARE and FARE equations. If the system is in the open-loop balanced representation, then the system is LQG balanced, with the Hankel singular values and the LQG singular values identical, $S_c = S_e = M = \Gamma^2$.

8.3 Balanced LQG Controller for Flexible Structures

Due to the specific properties of flexible structures the analytical properties of the LQG controller in this case can be readily exposed. First, it will be shown that flexible structures in the open-loop balanced representation produce diagonally dominant solutions of the CARE and FARE equations, and consequently an approximately LQG balanced representation. Assume a diagonal weight matrix $Q = diag(q_i I_2)$, $i = 1,...,n$, then there exist $q_i \leq q_{oi}$, where $q_{oi} > 0$, $i = 1,...,n$, such that $S_c \cong diag(s_{ci} I_2)$ is the solution of Eq.(8.3a), where

$$s_{ci} \cong \frac{\beta_{ci} - 1}{2\gamma_i^2}, \qquad \beta_{ci} = \sqrt{1 + \frac{2q_i \gamma_i^2}{\zeta_i \omega_i}} \qquad (8.18)$$

In order to prove it, note from Chapter 3 that for a balanced flexible

166

system (A,B,C) with n_2 components (or $n=2n_2$ states), the open-loop balanced matrix A is almost block-diagonal, with dominant 2×2 blocks on the main diagonal in modal form 1 or 2. Introducing A in modal form 2, and the property of Eq.(3.61), it follows immediately for a flexible structure that the off-diagonal terms of the solution S_c of the Riccati equation (8.3) are finite for $\zeta \to 0$, and the diagonal terms are proportional to $1/\zeta$, as $\zeta \to 0$. Moreover, the difference between the two diagonal entries of each pair is finite, hence S_c is diagonally dominant in the following form: $S_c \cong diag(s_{c1}, s_{c1}, \dots, s_{cn}, s_{cn})$. In this case Eq.(8.3) turns into a set of the following equations

$$s_{ci}(A_i + A_i^T) - s_{ci}^2 B_i B_i^T + q_i I_2 \cong 0, \qquad i=1,\dots,n. \tag{8.19}$$

For a balanced system $B_i B_i^T \cong -\gamma_i^2(A_i + A_i^T)$, and $A_i + A_i^T = -2\zeta_i \omega_i I_2$, see Eq.(3.61). Therefore Eq.(8.19) is now

$$s_{ci}^2 + \frac{s_{ci}}{\gamma_i^2} - \frac{q_i}{2\zeta_i \omega_i \gamma_i^2} \cong 0, \qquad i=1,\dots,n. \tag{8.20}$$

There are two solutions of the above equation, but for a stable system and for $q_i = 0$ it is required that $s_{ci} = 0$, therefore Eq.(8.18) is the unique solution of Eq.(8.20).

A similar result is obtained for the FARE equation. Namely, for a diagonal $V = diag(v_i I_2)$, $i=1,\dots,n$, there exist $v_i \leq v_{oi}$ where $v_{oi} > 0$, $i=1,\dots,n$, such that $S_e \cong diag(s_{ei} I_2)$ is the solution of Eq.(8.3b), where

$$s_{ei} \cong \frac{\beta_{ei} - 1}{2\gamma_i^2}, \quad \text{where} \quad \beta_{ei} = \sqrt{1 + \frac{2v_i \gamma_i^2}{\zeta_i \omega_i}} \tag{8.21}$$

For the diagonally dominant solutions of CARE and FARE, $(S_c \cong diag(s_{ci} I_2)$, $S_e \cong diag(s_{ei} I_2)$, $i=1,2,\dots,n)$, it is not difficult to find an approximately balanced solution M of CARE and FARE, which is

167

also diagonally dominant, i.e.,

$$M \cong diag(\mu_i I_2), \quad \text{where} \quad \mu_i = \sqrt{s_{ci} s_{ei}}, \quad i=1,\ldots,n \qquad (8.22)$$

The transformation R_{lqg} from the open-loop balanced representation (A_b, B_b, C_b) to the LQG balanced representation $(A_{lqg}, B_{lqg}, C_{lqg})$ is diagonally dominant as well,

$$R_{lqg} \cong diag(r_{lqg1} I_2, \ r_{lqg2} I_2, \ \ldots, \ r_{lqgn} I_2), \qquad r_{lqgi} = \left(\frac{s_{ei}}{s_{ci}}\right)^{1/4} \qquad (8.23)$$

where the LQG balanced representation is approximately obtained from the following transformation, $(A_{lqg}, B_{lqg}, C_{lqg}) \cong (A_b, R_{lqg}^{-1} B_b, C_b R_{lqg})$. Note that this transformation requires only a re-scaling of the components.

The diagonally dominant solutions of CARE and FARE allow one to determine the relationship between the weights and the root locus, which is a useful tool of controller design. Let the weight Q be

$$Q = diag(0,0,\ldots,q_i I_2,\ldots 0,0), \qquad (8.24a)$$

then for $q_i \leq q_{oi}$ the closed-loop pair of flexible poles $(\lambda_{cri}, \pm j\lambda_{cii})$ relates to the open-loop poles $(\lambda_{ori}, \pm j\lambda_{oii})$ as follows

$$(\lambda_{cri}, \pm j\lambda_{cii}) \cong (\beta_{ci}\lambda_{ori}, \pm j\lambda_{oii}), \qquad i=1,\ldots,n \qquad (8.25a)$$

where β_{ci} is defined in Eq.(8.18). In order to prove it, note that for small weight q_i the matrix A of the closed-loop system is diagonally dominant, i.e., $A_o \cong diag(A_{oi})$, $i=1,\ldots,n$, and $A_{oi} = A_i - B_i B_i^T s_{ci}$. Introducing Eq.(3.60), one obtains

$$A_{oi} \cong A_i + 2s_{ci} r_i^2 (A_i + A_i^T), \qquad (8.26)$$

and introducing A_i as in Eq.(A.2) to Eq.(8.26), one obtains

$$A_{oi} \cong \begin{bmatrix} -\beta_{ci}\zeta_i\omega_i & -\omega_i \\ \omega_i & -\beta_{ci}\zeta_i\omega_i \end{bmatrix}, \tag{8.27}$$

with β_{ci} as in Eq.(8.18).

This result implies that the weight Q as in Eq.(8.24a) shifts the ith pair of complex poles of the flexible structure, and leaves the remaining pairs of poles almost unchanged. Only the real part of the pair of poles is changed (just moving the pole apart from the imaginary axis and stabilizing the system), and the imaginary part of the poles remains unchanged.

The above proposition has additional interpretations. Note that the real part of the ith open-loop pole is $\lambda_{oi} = -\zeta_i\omega_i$, and that the real part of the ith closed-loop pole is $\lambda_{ci} = -\zeta_{ci}\omega_i$; note also that the height of the open-loop resonant peak is $\alpha_{oi} = \kappa/2\zeta_i\omega_i$, where κ is a constant, and the closed-loop resonant peak is $\alpha_{ci} = \kappa/2\zeta_{ci}\omega_i$. From Eq.(8.25a) one obtains $\beta_{ci} = \lambda_{cri}/\lambda_{ori}$, hence

$$\beta_{ci} = \frac{\zeta_{ci}}{\zeta_i} = \frac{\alpha_{oi}}{\alpha_{ci}} \tag{8.28}$$

is a ratio of the closed- and open-loop damping factors, or it is a ratio of the open- and closed-loop resonant peaks. Therefore, if a suppression of the ith resonant peak by the factor β_{ci} is required, the appropriate weight q_i is determined from Eq.(8.18)

$$q_i \cong \frac{(\beta_{ci}^2 - 1)\zeta_i\omega_i}{2\gamma_i^2} \tag{8.29}$$

The coefficient β_{ci} can be also interpreted as a ratio of the open- and closed-loop Hankel singular values, or as a ratio of the variances of open-loop (σ_{oi}^2) and closed-loop (σ_{ci}^2) states excited by the white noise input

$$\beta_{ci} \cong \frac{\gamma_{oi}^2}{\gamma_{ci}^2} = \frac{\sigma_{oi}^2}{\sigma_{ci}^2} \qquad (8.30)$$

This interpretation follows from the closed-loop Lyapunov equation

$$(A-BB^TS_c)\Gamma_c^2 + \Gamma_c^2(A-BB^TS_c)^T + BB^T = 0 \qquad (8.31a)$$

which for the ith pair of variables is as follows

$$(A_i - B_i B_i^T S_{ci})\gamma_{ci}^2 + \gamma_{ci}^2(A_i - B_i B_i^T S_{ci})^T + B_i B_i^T \cong 0 \qquad (8.31b)$$

Introducing Eq.(3.61) gives

$$\gamma_{ci}^2 + 2\gamma_{ci}^2\gamma_{oi}^2 s_{ci} - \gamma_{oi}^2 \cong 0 \qquad (8.32)$$

or

$$\frac{\gamma_{oi}^2}{\gamma_{ci}^2} \cong 1 + 2s_{ci}\gamma_{oi}^2 = \beta_{ci} \qquad (8.33)$$

The plots of β_{ci} with respect to the weight q_i and the Hankel singular value γ_i are shown in Fig.8.2a,b.

The estimator poles are shifted in a similar manner. Denote

$$V = diag(0,0,\ldots,v_iI_2,\ldots0,0), \qquad (8.24b)$$

then $v_i \leq v_{oi}$, and the estimator pair of poles $(\lambda_{eri},\pm j\lambda_{eii})$ relates to the open-loop poles $(\lambda_{ori},\pm j\lambda_{oii})$ as follows

$$(\lambda_{eri},\pm j\lambda_{eii}) \cong (\beta_{ei}\lambda_{ori},\pm j\lambda_{oii}), \qquad i=1,\ldots,n \qquad (8.25b)$$

where β_{ei} is defined in Eq.(8.21).

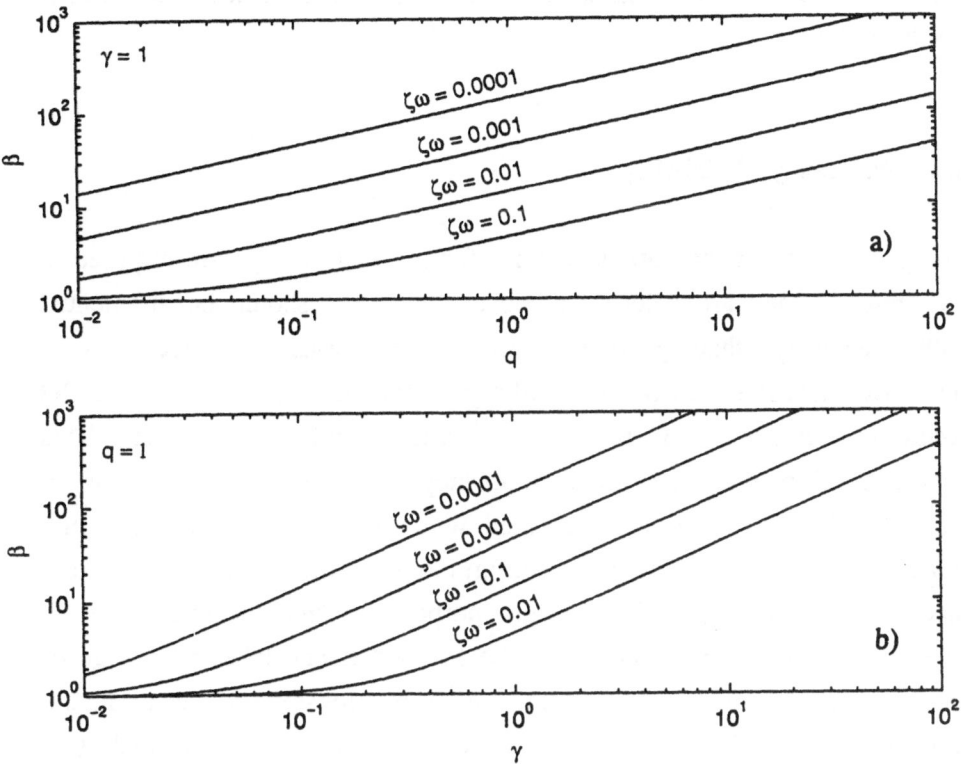

Figure 8.2: Coefficient β vs. (a) weight, and (b) Hankel singular value

The above applies for controllers with the limited damping authority, i.e. controllers that modify only moderately the system natural frequencies, as defined by Aubrun and Marguiles, see Refs.[5] and [6]. The controller authority is limited by the values q_{oi} and v_{oi} introduced earlier. For limited damping authority one has $q_i \leq q_{oi}$, and $v_i \leq v_{oi}$. The limiting values q_{oi}, and v_{oi} are not difficult to determine. There are several symptoms that the weight q_i is approaching q_{oi} (or that v_i is approaching v_{oi}). Namely, q_{oi} is the weight at which the ith pair of complex poles of the plant departs significantly from the horizontal trajectory in the root-locus plane and approaches the real axis, or it is the weight at which the ith resonant peak of the plant transfer function disappears (the peak is flattened). A similar result applies to v_{oi}.

8.4 Frequency Weighting

The process and/or measurement noise for the LQG design are required to be white. If they are colored, however, the plant should be augmented with frequency shaping filters. Also, the noises are frequently a fictitious variables, used as auxiliary tools in the design of LQG controllers rather than to reflect "real world" phenomena. This approach often will call for a frequency dependent weighting, as in Ref.[3], where, a plant is augmented with input and/or output shaping filters, see Fig.8.3. But this configuration forms a modified plant model used for the LQG design purposes. This inconvenience of implementation of a new model can be avoided if one takes advantage of the already known (from Section 4.8.1) relationship between Hankel singular values between the augmented (γ_{ai}^2) and the original (γ_i^2) plant. Let (A,B,C) be a balanced representation of a plant, and consider for conciseness an input filter only, with the transfer function W_{in}. Then

$$\gamma_{ai}^2 = \alpha_i \gamma_i^2, \qquad i=1,\dots,n_2 \tag{8.34a}$$

where

$$\alpha_i = w_{ini} \, \cos \, \phi_i, \tag{8.34b}$$

$$w_{ini} = \| W_{in}(\omega_i) \|, \qquad \cos \, \phi_i = \frac{\| c_i b_i W_{in}(\omega_i) \|}{w_{ini} \| c_i b_i \|} \tag{8.34c}$$

and c_i, b_i are the ith column of C, and the ith row of B, respectively.

172

Thus, frequency weighting in the LQG controller design procedure from the Section 8.3 is achieved by the replacement of γ_i with γ_{ai}.

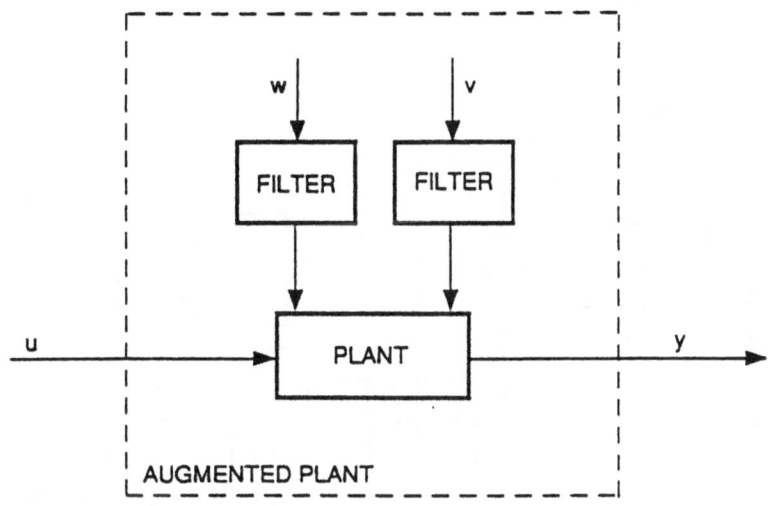

Figure 8.3: Flexible structure with shaping filters

8.5 Tracking LQG Controller

The LQG controllers for flexible structures as considered previously are designed for vibration suppression due to external disturbances. A more complex task includes a tracking controller with a required tracking performance in addition to vibration suppression properties. This is the case of controllers for radars and microwave antennas, such as the NASA Deep Space Network antennas. A configuration for this kind of controller is different than the one in Fig.8.1. Namely, in order to assure zero steady-state tracking error, an integral of the plant position should be added to the plant state-space representation, as reported in Refs.[4], [28], [29], [58], [96], [121], and [122]. Thus, the closed-loop system with tracking LQG controller is shown in Fig.8.4, with the plant state-space triple (A,B,C), the state x, estimated state $\hat{x}$, estimated state of flexible part $\hat{x}_f$, command r, control input u, output y, estimated output $\hat{y}$, servo error $e=r\text{-}\hat{y}$,

173

integral of servo error e_i, process noise v of intensity V, and measurement noise w of intensity W, with both v and w uncorrelated: $V=E(vv^T)$, $W=E(ww^T)=I$, $E(vw^T)=0$, $E(v)=0$, $E(w)=0$.

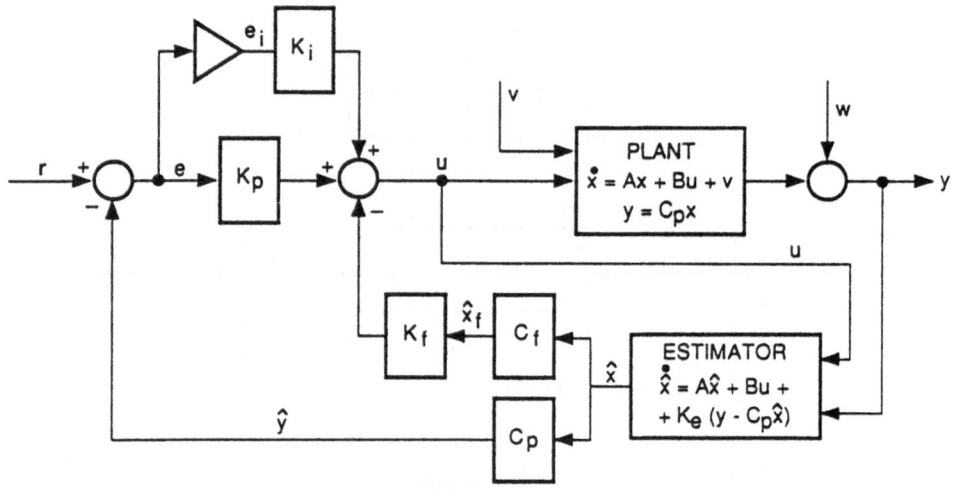

Figure 8.4: Tracking LQG controller

For the open-loop state-space representation (A,B,C) of a flexible structure divide the state vector x into the tracking x_t and flexible x_f parts

$$x=\begin{bmatrix} x_t \\ x_f \end{bmatrix} \tag{8.35}$$

For the state as above the system triple can be presented as follows (see Ref.[43])

$$A=\begin{bmatrix} A_t & A_{tf} \\ 0 & A_f \end{bmatrix}, \quad B=\begin{bmatrix} B_t \\ B_f \end{bmatrix}, \quad C=[C_t \ \ 0] \tag{8.36a}$$

The gain, K_c, the weight, Q, and the solution of CARE, S_c, are divided similarly to x

174

$$K_c = [K_{ct} \quad K_{cf}], \qquad Q = \begin{bmatrix} Q_t & 0 \\ 0 & Q_f \end{bmatrix}, \qquad \text{and} \qquad S_c = \begin{bmatrix} S_{ct} & S_{ctf} \\ S^T_{ctf} & S_{cf} \end{bmatrix} \qquad (8.36b)$$

The tracking system is considered to be a low-authority system, if the flexible weights are much smaller than the tracking ones, i.e., such that $\|Q_t\| \gg \|Q_f\|$. It was shown by Collins $et\ al.$, in Ref.[17] that for $Q_f = 0$ one obtains $S_{cf} = 0$ and $S_{ctf} = 0$. It means that the gain of the tracking part, K_{ct}, does not depend on the flexible part. And for the low-authority tracking system (with small Q_f), one obtains weak dependence of the tracking gains on the flexible weights, due to the continuity of the solution. A similar conclusions apply to the FARE equation (8.3b).

This property can be validated by the observation of the closed-loop transfer functions for different weights for the Deep Space Network antenna, as in Fig.8.5. Denote I_n and 0_n the identity and zero matrix of order n, then the magnitude of the closed-loop transfer function (elevation angle-to-elevation command) for $Q_t = I_4$ and $Q_f = 0_{26}$ is shown as a solid line, for $Q_t = I_4$ and $Q_f = 10^{-7} \times I_{26}$ as a dashed line, and for $Q_t = 20 \times I_4$ and $Q_f = 0_{26}$ as a dot-dashed line. It follows from the plots that variations in Q_f changed the properties of the flexible subsystem only, while variations in Q_t changed the properties of both subsystems.

Note, however, that the larger Q_f increases dependency of the gains on the flexible system; only a quasi-independence in the final stage of controller design is observed, while the separation in the initial stages of controller design is still strong. The design consists therefore of the initial choice of weights for the tracking subsystem, and determination of the controller gains of the flexible subsystem. It is followed by the adjustment of weights of the tracking subsystem, and a final tuning of the flexible weights, if necessary.

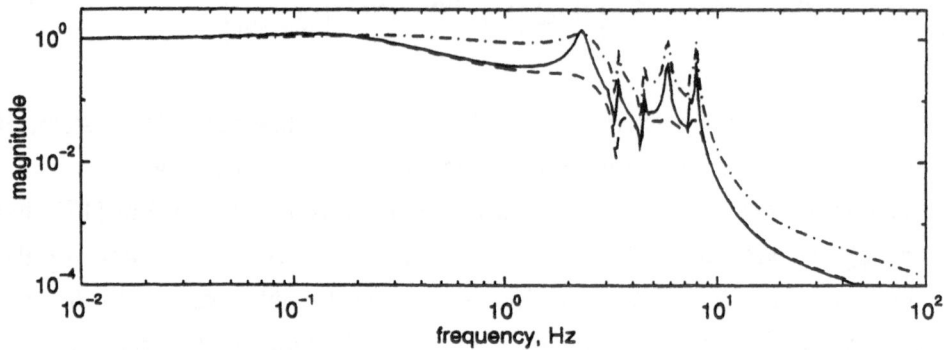

Figure 8.5: Magnitude of transfer function of closed-loop system for different LQG weights

This approach is similar to the high-/low-authority controller design, reported in Refs.[5], [6], and [110]. High-authority controller is designed to assure required performance, while low-authority controller is used as a robust inner loop for suppressing the plant responses outside the high-authority controller bandwidth. The low-authority controller has limited damping authority, i.e.,   it is allowed to modify only moderately the system poles, see Ref.[6].

8.6 Reduced-Order LQG Controller

Although the size of a plant determines the size of a controller, this size is most often not acceptable from an implementation point of view. It is crucial to obtain a controller of the smallest possible order that preserves the stability and performance of the full-order controller. But in order to assure the quality of the closed-loop system design, the plant model should not be reduced excessively in advance. Therefore, controller reduction is a part of controller design. The balanced LQG design procedure provides this opportunity.

8.6.1 Reduction Index

In order to successfully perform the controller reduction, an index of the importance of each controller component is introduced. In the open-loop case, Hankel singular values serve as reduction indices. In the closed-loop case, the LQG characteristic values were used as reduction indices by Jonckheere and Silverman, Ref.[60] for symmetric and passive systems, but, unfortunately, they produce unstable controllers for other systems.

Here, the effectiveness of the closed-loop system is evaluated by the degree of suppression of flexible motion of the structure. The suppression, in turn, depends on the pole mobility to the right-hand side of the complex plane. Therefore, if a particular pair of poles is easily moved (i.e., when a small amount of weight is required to move the poles), the respective states are easy to control and to estimate. On the contrary, if a particular pair of poles is difficult to move (i.e., even a large amount of weight insignificantly moves the poles), the respective states are difficult to control and to estimate. In the latter case, the action of the controller is irrelevant, and the states which are difficult to control and to estimate can be reduced; this demonstrates the rationale of the choice of the pole mobility as an indicator of the importance of controller states.

For the LQG balanced system the reduction index σ_i is defined as a product of the square of a Hankel singular value and the LQG characteristic value of a system

$$\sigma_i = \gamma_i^2 \mu_i \tag{8.37a}$$

It combines the observability and controllability properties of the open-loop system and the controller performance. This choice is not an accidental one, but is a result of the fact that σ_i is a measure of the ith pole mobility. In order to show that note from Eqs.(8.18), (8.21),

177

(8.22), and (8.37a) that σ_i the geometric mean of the plant and the estimator pole mobility indexes, i.e.

$$\sigma_i \cong 0.5\sqrt{(\beta_{ci}-1)(\beta_{ei}-1)} \qquad (8.37b)$$

This equation reveals, for example, that for $\beta_{ci}=1$ the ith controller pole is stationary, and σ_i is equal to zero. Similarly, for $\beta_{ei}=1$ the ith estimator pole is stationary, and σ_i is equal to zero. However, for a shifted pole one obtains $\beta_{ci}>1$, $\beta_{ei}>1$, hence the index is also "shifted," that is $\sigma_i>0$.

Denote σ_{oi}^2 the variance of the open-loop response to white noise, and σ_{ci}^2 the variance of the closed-loop response to white noise, and note that $\gamma_{oi}^2=\sigma_{oi}^2$, $\gamma_{ci}^2=\sigma_{ci}^2$, where γ_o, and γ_c are the open- and closed-loop Hankel singular values. Denote also $\Delta\gamma_i^2=\gamma_{oi}^2-\gamma_{ci}^2$, and $\Delta\sigma_i^2=\sigma_{oi}^2-\sigma_{ci}^2$ the change of the response of the open- and the closed-loop systems due to white noise. Then another useful interpretation of the reduction index follows from Eq.(8.33):

$$\sigma_i \cong \frac{\Delta\gamma_i^2}{2\gamma_{ci}^2} \cong \frac{\Delta\sigma_i^2}{2\sigma_{ci}^2} \qquad (8.38)$$

This equation shows that the reduction index is proportional to the relative change of the response of the open- and the closed-loop systems due to white noise. Having defined σ_i as the controller performance evaluation tool, the reduction technique is developed.

8.6.2 Reduction Technique

In order to introduce the controller reduction technique the matrix Σ of reduction indices is defined $\Sigma=diag(\sigma_1, \sigma_2, ..., \sigma_n)$, and from Eq.(8.37a) it follows immediately that

$$\Sigma = \Gamma^2 M \qquad (8.39)$$

Next, the diagonal entries σ_i in Σ are put in the descending order, i.e., $\sigma_i \geq 0$, $\sigma_i \geq \sigma_{i+1}$, $i = 1, \ldots, n$, and the matrix is divided it as follows

$$\Sigma = \begin{bmatrix} \Sigma_r & 0 \\ 0 & \Sigma_t \end{bmatrix}, \qquad (8.40)$$

where Σ_r consists of first k entries of Σ, and Σ_t the remaining ones. If the entries of Σ_t are small in comparison with the entries of Σ_r, the controller can be reduced by truncating its last n-k states.

Note that the value of the index σ_i depends on the weight q_i, so that the reduction depends on the weight choice. For example, if for a given weight a particular resonant peak is too large to be accepted (or a pair of poles is too close to the imaginary axis, or amplitudes of vibrations at this resonance frequency are unacceptable high), the weighting of this particular component should be increased to suppress this component. The growth of weight increases the value of σ_i, which can save this particular component from reduction.

8.6.3 Stability of the Reduced-Order Controller

There is no doubt that the question of stability of the closed-loop system with the reduced-order controller should be answered before implementation. In order to answer this fundamental question, consider the closed-loop system as in Fig.8.1. Denoting the state $x_o = [x^T \ \varepsilon^T]^T$, and $\varepsilon = x - \hat{x}$, one obtains the following closed-loop equations

$$\dot{x}_o = A_\sigma x_o + B_o r + B_v v + B_w w, \qquad y = C_\sigma x_o + w \qquad (8.41a)$$

where

$$A_o = \begin{bmatrix} A\text{-}BK_p & BK_p \\ 0 & A\text{-}K_e C \end{bmatrix}, \qquad (8.41b)$$

$$B_o = \begin{bmatrix} B \\ 0 \end{bmatrix}, \quad B_v = \begin{bmatrix} I \\ I \end{bmatrix}, \quad B_w = \begin{bmatrix} 0 \\ -K_e \end{bmatrix}, \quad C_o = [C \ \ 0] \qquad (8.41c)$$

Let the matrices A, B, C of the estimator (in the lower row and the right column of the above equation) be partitioned conforming to Σ in Eq.(8.40)

$$A \cong \begin{bmatrix} A_r & 0 \\ 0 & A_t \end{bmatrix}, \quad B = \begin{bmatrix} B_r \\ B_t \end{bmatrix}, \quad C = [C_r \ \ C_t] \qquad (8.42)$$

then the reduced controller representation is (A_r, B_r, C_r). The controller gains are divided similarly

$$K_p = [K_{pr} \ \ K_{pt}], \qquad K_e = \begin{bmatrix} K_{er} \\ K_{et} \end{bmatrix} \qquad (8.43)$$

and the resulting reduced closed-loop system is as follows

$$A_{or} = \begin{bmatrix} A\text{-}BK_p & BK_{pr} \\ 0 & A_r\text{-}K_{er} C_r \end{bmatrix}, \qquad (8.44a)$$

$$B_{or} = \begin{bmatrix} B \\ 0 \end{bmatrix}, \quad B_{vr} = \begin{bmatrix} B \\ B_r \end{bmatrix}, \quad B_{wr} = \begin{bmatrix} 0 \\ -K_{er} \end{bmatrix}, \quad C_{or} = [C \ \ 0_r] \qquad (8.44b)$$

Define the stability margin of matrix A_o as follows

$$m(A_o) = \min_i(-Re(\lambda_i(A_o))) \qquad (8.45)$$

where $Re(.)$ denotes a real part of a complex variable, and $\lambda_i(.)$ is the

180

*i*th eigenvalue of a matrix. For $\|\Sigma_t\| \ll \|\Sigma_r\|$ and $m(A_v) \geq m(A_o)$, one obtains $m(A_o) \cong m(A_{or})$, where A_{or} is a closed-loop matrix of a system with the reduced controller, and A_t is the state matrix of the truncated part.

In order to prove it, introduce Eqs.(8.42) and (8.43) to Eq.(8.41b) to obtain

$$
A_o = \left[
\begin{array}{cccc}
A_r\text{-}B_rK_{pr} & -B_rK_{pt} & B_rK_{pr} & B_rK_{pt} \\
-B_tK_{pr} & A_t\text{-}B_tK_{pt} & B_tK_{pr} & B_tK_{pt} \\
0 & 0 & A_r\text{-}K_{er}C_r & -K_{er}C_t \\
\hline
0 & 0 & -K_{et}C_r & A_t\text{-}K_{et}C_t
\end{array}
\right]
\tag{8.46}
$$

the matrix A_o divided into four blocks. The upper left block of A_o is A_{or}, thus in order to prove that $m(A_o) \cong m(A_{or})$, it is sufficient to show that: (a) in the lower left block $\|K_{et}C_r\| \cong 0$, and (b) $m(A_v) \geq m(A_o)$, i.e., that in the lower right block $\|K_{et}C_t\| \cong 0$. But for (a), from Eq.(3.61) one obtains $\|K_{et}C_r\| = \|M_t C_t C_r\| \ll \|M C_t C_r\| \cong \|\Sigma_t\| \cong 0$; similarly, for (b) $\|K_{et}C_t\| = \|M C_t C_t\| \cong \|\Sigma_t\| \cong 0$.

8.6.4 Performance of the Reduced-Order Controller

In addition to the stability evaluation, the performance of the reduced-order controller is assessed. Denote $\varepsilon^T = [\varepsilon_r^T, \ \varepsilon_t^T]$ the estimation error of the full-order controller and ε_{rr} the estimation error of the reduced-order controller. If the states with small reduction indices are truncated then $\|\Sigma_t\| \ll \|\Sigma_r\|$, and one obtains

$$
\varepsilon_r \cong \varepsilon_{rr} \text{ and } \varepsilon_t \cong 0,
\tag{8.47}
$$

In order to prove it, note that for A_o as in Eq.(8.46) the estimation error is

$$
\dot\varepsilon_r = (A_r\text{-}K_{er}C_r)\varepsilon_r\text{-}K_{er}C_t\varepsilon_t, \quad \dot\varepsilon_t = -K_{et}C_r\varepsilon_r + (A_t\text{-}K_{et}C_t)\varepsilon_t
\tag{8.48a}
$$

But, from Eq.(8.44) the error of the reduced-order controller is

$$\dot{\varepsilon}_{rr} = (A_r - K_{er} C_r)\varepsilon_{rr} \qquad (8.48b)$$

It was shown previously that $\|K_{er} C_t\| \cong 0$, and $\|K_{et} C_t\| \cong 0$ for small σ_i, thus $\varepsilon_r \cong \varepsilon_{rr}$. Additionally one obtains $\dot{\varepsilon}_t \cong A_t \varepsilon_t$, imposing that for stable A_t the truncation error vanishes ($\varepsilon_t \to 0$) with time elapsing, ($t \to \infty$).

The above property implies that for $\|\Sigma_t\| \ll \|\Sigma_r\|$ the performance of the reduced- and full-order controllers is approximately the same. The performance of full- and reduced-order controllers is compared in the design examples section.

8.7 Procedure for the Balanced LQG Controller Design

1. Balance (in the Moore sense) the plant state-space representation.
2. Chose the matrices Q and V to obtain the required performance (this is determined either through pole placement, see Eq.(8.29), or through observation of resonant peaks of the closed-loop system, or through the time-domain performance, e.g., step or impulse responses).
3. Find the LQG balanced solution M of CARE and FARE, along with controller and estimator gains.
4. Form the reduction matrix Σ as in Eqs.(8.37) and (8.39). If some diagonal entries of Σ are much smaller than the remaining ones, truncate the estimator states related to these entries.
5. Check stability and performance of the reduced-order controller.

8.8 Design Examples

Three illustrations of a balanced LQG controller design are presented: (1) for the simple flexible system, (2) for the truss structure, and (3) for the Deep Space Network antenna.

182

8.8.1 Simple Structure

The system is shown in Fig.2.1, with masses $m_1=m_2=m_3=1$, stiffness $k_1=10, k_2=3, k_3=4, k_4=0$ and a damping matrix $D=0.004K+0.001M$, where K, M are stiffness and mass matrices, respectively. The input force is applied to the mass m_3, the output is the rate of the same mass, and the poles of the open-loop system are

$$\lambda_{o1,o2}=-0.0024\pm j0.9851,$$
$$\lambda_{o3,o4}=-0.0175\pm j2.9197,$$
$$\lambda_{o5,o6}=-0.0295\pm j3.8084.$$

The system open-loop balanced representation was obtained, with the Hankel matrix of the plant

$$\Gamma=diag(7.9776,\ 7.9776,\ 2.2337,\ 2.3336,\ 0.4893, 0.4890)$$

The weight matrix Q and the covariance matrix V are chosen as follows: $Q=V=diag(0.4,\ 0.4,\ 2,\ 2,\ 6,\ 6)$. For these matrices the solution S_c of CARE and the solution S_e of FARE are equal and diagonally dominant, i.e., it is approximately LQG balanced

$$S_c=S_e\cong M=diag(1.3288,\ 1.3261,\ 4.3161,\ 4.1301,\ 25.2817,\ 24.0490),$$

The corresponding gains are sign-symmetric, see Ref.[60]

$$k_p=[0.0039,\ 0.8893,\ 0.2978,\ -1.9291,\ 2.1378,\ -2.1636]$$
$$k_e^T=[-0.0039,\ 0.8893,\ -0.2978,\ -1.9291,\ -2.1378,\ -2.1636]$$

and the reduction matrix Σ is obtained

$$\Sigma\cong diag(84.5658,\ 84.3920,\ 21.5332,\ 20.6058,\ 6.0453,\ 5.7586)$$

The nonzero entries of Q and V shift the poles to the right, so that the peaks of the open-loop transfer function (solid line in Fig.8.6) are flattened in the closed-loop transfer function (dashed line in

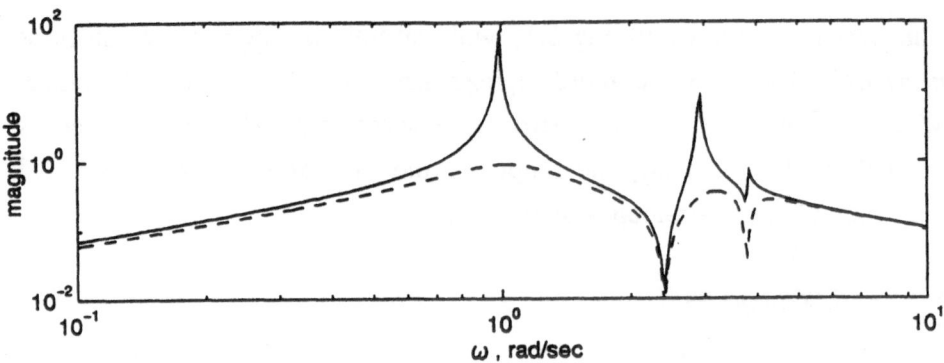

Figure 8.6: Magnitudes of transfer function of open- and closed-loop simple flexible system

Fig.8.6). Poles of the open-loop plant, closed-loop systems, and estimator are shown in Fig.8.7. The closed-loop poles and estimator poles were shifted horizontally with respect to the open-loop poles, in agreement with Eq.(8.25). For the chosen weights the projected (from Eq.(8.25)) and actual shifts are 137 vs 146 for the first pair of

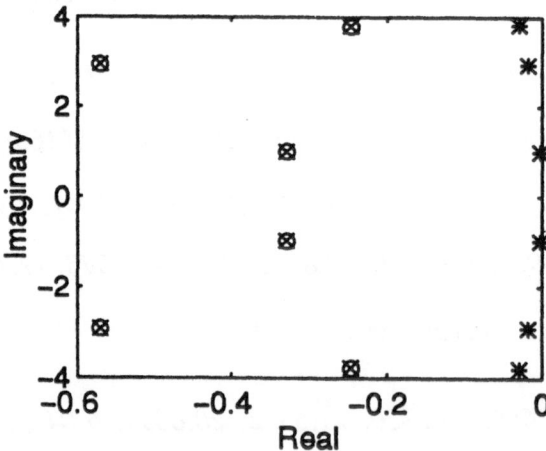

Figure 8.7: Poles of open- (*) and closed-loop (∘) simple flexible system, and poles of the estimator (×)

poles, 33 vs 34 for the second pair of poles, and 8.5 vs 10 for the third pair of poles. Moreover, since $Q=V$, the poles of the closed-loop system and the estimator overlap. The open-loop (solid line) and closed-loop (dashed line) impulse responses are shown in Fig.8.8, showing good vibration damping properties for the closed-loop system, which is also confirmed by the closed-loop transfer function (Fig.8.6, dashed line). The controller is reduced from six to four state variables. The truncated states are related to the smallest diagonal entries *(5.7586, 6.0453)* of Σ. The impulse responses of the full- (solid line) and reduced-order (dashed line) controller are compared in Fig.8.9, showing good coincidence. However, if the two states corresponding to the medium values of Σ *(20.6058, 21.5332)* are deleted, while the remaining four states are retained, the performance of the reduced-order controller is significantly deteriorated. And if the states corresponding to the largest entries of Σ are reduced, while the remaining four states are retained, the controller is unstable.

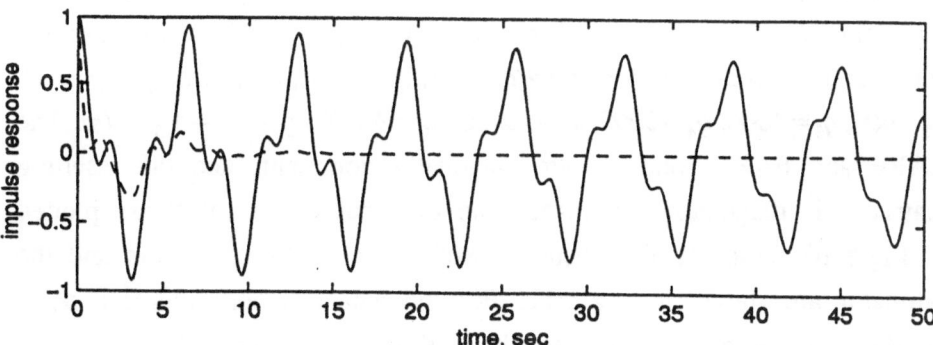

Figure 8.8: Impulse response of open- and closed-loop simple flexible system

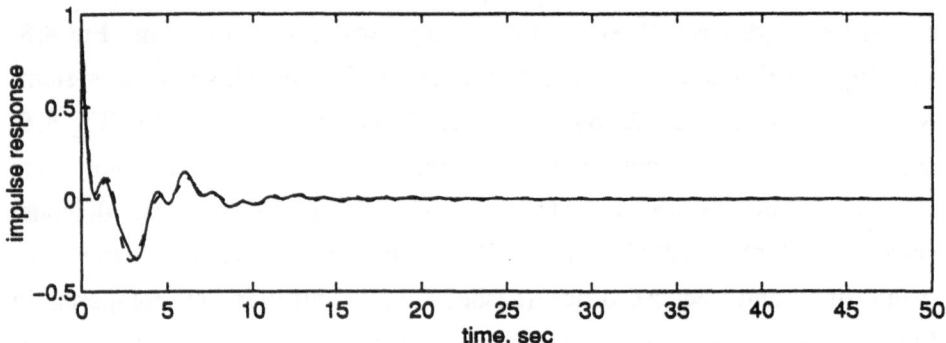

Figure 8.9: Impulse responses of full- and reduced-order controller for simple flexible system

8.8.2 Truss

The truss is presented in Fig.2.2. Vertical control forces are applied at nodes $n7$ and $n8$, and the output rates are measured in the vertical direction at nodes $n3$ and $n4$. The system has 26 states (13 balanced components), two inputs, and two outputs. For the open-loop balanced representation the weight (Q) and covariance (V) matrices are assumed equal and diagonal, $Q=V=diag(q_1,q_1,q_2,q_3,\ldots,q_{13},q_{13})$, where $q_1=200$, $q_2=400$, $q_3=1000$, $q_4=2000$, $q_5=10000$, $q_6=20000$, $q_7=\ldots=q_{13}=200$. The CARE and FARE solutions are diagonally dominant, and the resulting matrix Σ is diagonally dominant. The diagonal entries of Σ are plotted in Fig.8.10. Poles of the open- as well the closed-loop system and the estimator are shown in Fig.8.11. For the balanced controller the poles of the closed-loop system and the estimator overlap. The open-loop (solid line) and closed-loop (dashed line) impulse responses from input $n7$ to output $n3$ are shown in Fig.8.12, showing good damping properties for the closed-loop system. The open-loop transfer function (solid line) is shown in Fig.8.13 and the closed-loop transfer functions

186

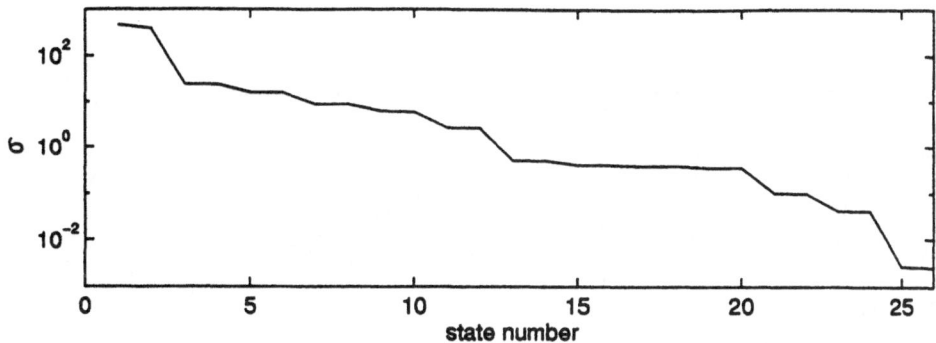

Figure 8.10: Reduction indices for the truss structure

(dashed line) of the plant and estimator overlap in Fig.8.13, and show that the oscillatory motion of the structure is damped out. The controller is reduced by truncating the 14 states, which correspond to the small reduction indices $\sigma_i < 0.5$. The resulting reduced-order controller has 12 states. The impulse responses of the full- (solid line) and reduced-order (dashed line) controller are compared in Fig.8.14, showing good coincidence.

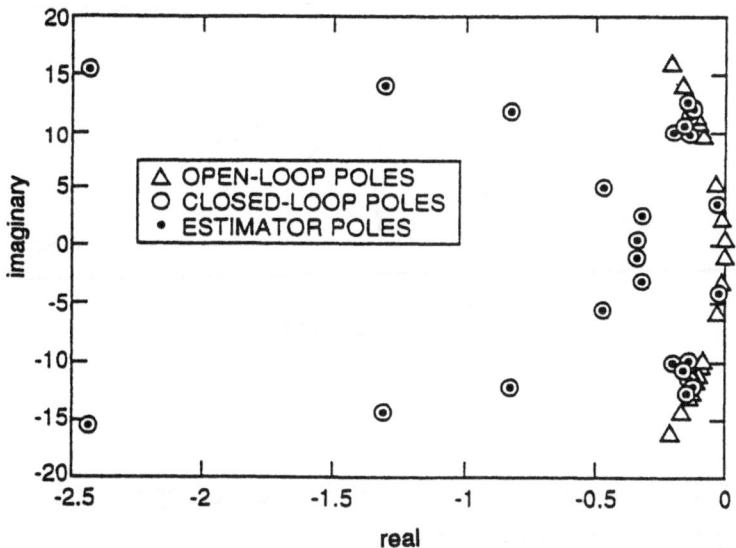

Figure 8.11: Poles of open- and closed-loop truss structure, and poles of the estimator

187

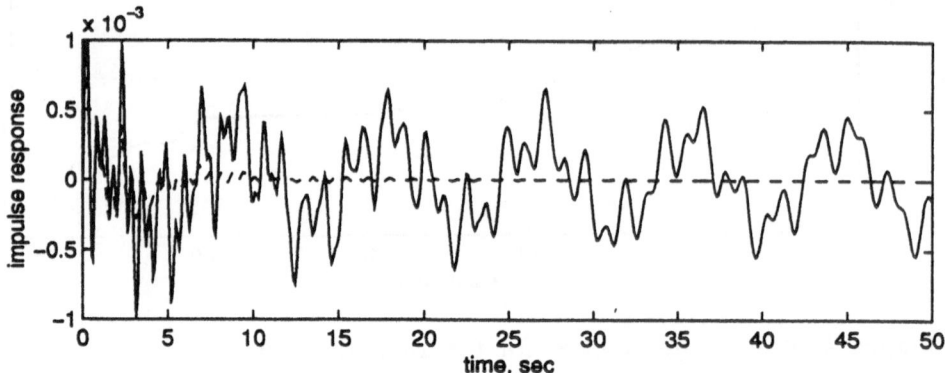

Figure 8.12: Impulse response of open- and closed-loop truss structure, at the first output

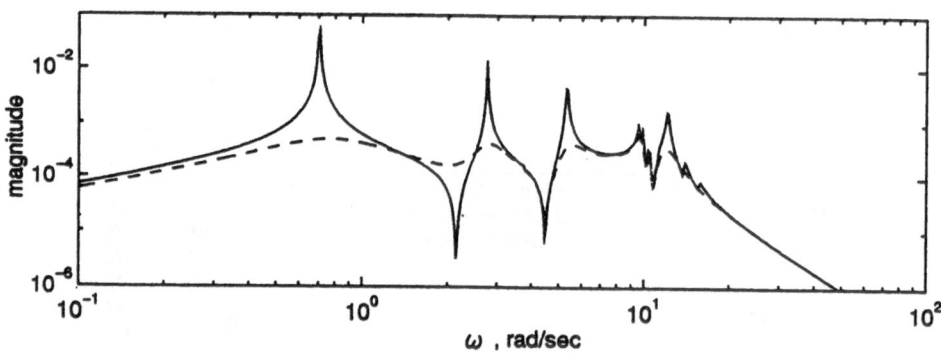

Figure 8.13: Magnitudes of transfer function of open- and closed-loop truss structure, at the first output

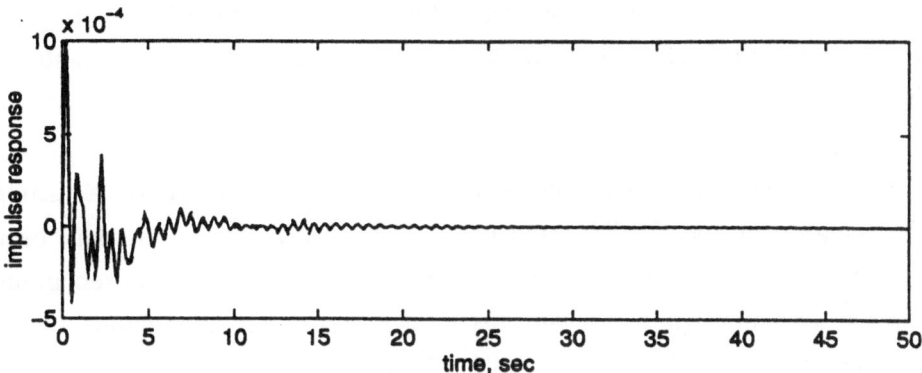

Figure 8.14: Impulse responses of full- and reduced-order controller for truss structure, at the first output

8.8.3 Deep Space Network Antenna

The design of a balanced LQG compensator is illustrated with a Deep Space Network antenna. For the open-loop balanced representation the weight, Q, and plant noise covariance, V, are assumed equal, therefore in the following only the determination of Q is discussed. The tracking of the trajectory as in Fig.8.15a was simulated. The maximal rate of this trajectory is 0.4 deg/sec, as expected for some of the low-orbiting spacecraft. The pointing errors are used as a measure of antenna performance, and the step responses and Bode plots are used as complimentary measures of performance.

The Deep Space Network antenna model consists of four tracking states (azimuth and elevation angles and their integrals) and 13 flexible modes (or 26 balanced states). The preliminary weights $q_{ie}=q_{pe}=q_{ia}=q_{pa}=1$ for the tracking subsystem (for the angles y and their integrals y_i) and zero weights for the flexible subsystem ($q_1=q_2=...=q_{13}=0$ for all 13 modes) were chosen. The closed-loop system

189

step response is presented in Fig.8.16a (azimuth encoder reading due to azimuth command), and the magnitudes of the closed-loop transfer function in Fig.8.16b, showing that the flexible motion of the antenna is excessive.

The excess of flexible deformations can be suppressed by adjusting the weights of the flexible subsystem. Thus, the weight 10^{-7} has been applied to the first 10 flexible modes, and weights for the remaining modes (11th through 13th) have been set to zero. The suppression of the flexible motion is observed in Fig.8.17a,b.

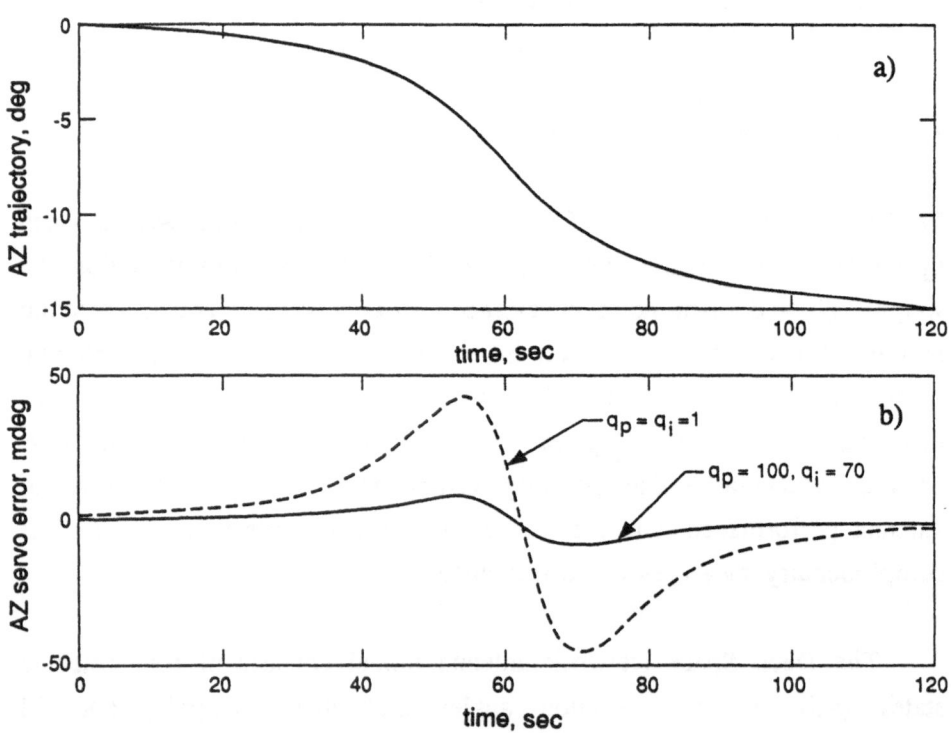

Figure 8.15: (a) Spacecraft trajectory, and (b) tracking error for different LQG weights

190

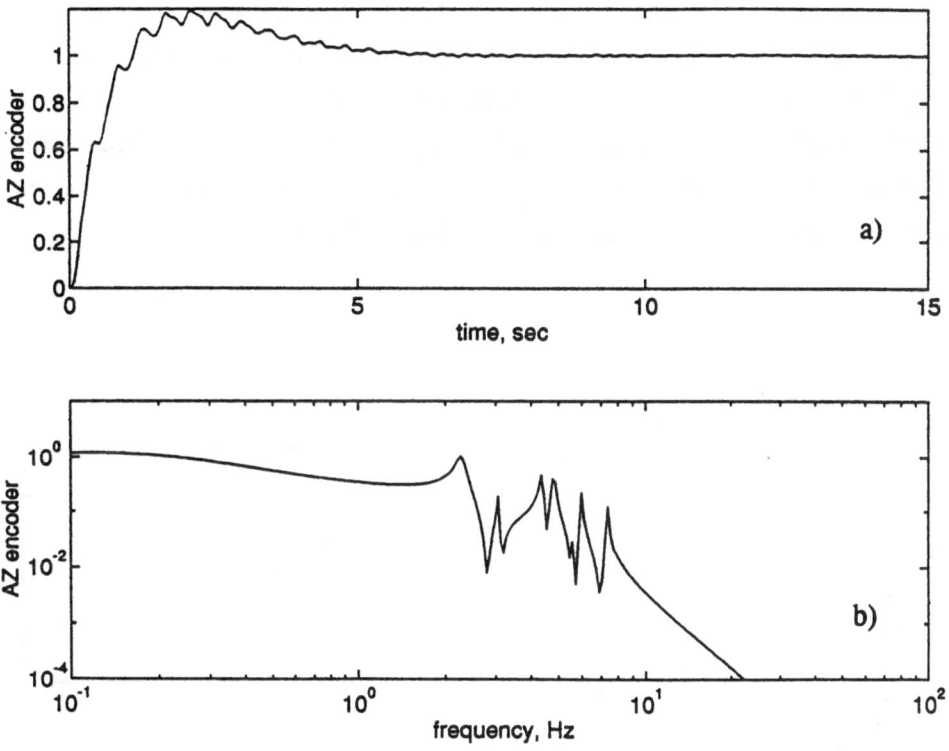

Figure 8.16: Antenna (a) step response, and (b) transfer function, for unit PI weights, and zero "flexible" weights

In the next step, the tracking properties of the system are improved by weight setting of the tracking subsystem. By setting the integral weight to $q_{ie}=q_{ia}=70$ and the proportional weight to $q_{pe}=q_{pa}=100$, the tracking properties are improved, as in Fig.8.18a (small overshoot and settling time) and in Fig.8.18b (extended bandwidth up to 2 Hz). However, the transfer function has been raised in the frequency region of 1-3 Hz, forcing the two modes located in this region to appear again in the step response. By sacrificing a bit

191

of the tracking properties, the flexible motion in the step response is removed by increasing the weights of the flexible subsystem, setting them as follows: $q_1=q_2=\ldots=q_6=10^{-6}$, $q_7=q_8=10^{-7}$, $q_9=q_{10}=10^{-5}$. The closed-loop system response with satisfactory tracking performance is shown in Figs.8.19a and 8.19b (small overshoot, settling time, and 1-Hz bandwidth). This is confirmed by comparing the tracking errors in Fig.8.15b (solid line) for the latter case, and for the case with the same flexible weights and tracking weights equal to 1 (dashed line). The maximal azimuth error has been reduced from 45 to 8 mdeg, while the elevation error has been reduced from 102 to 18 mdeg.

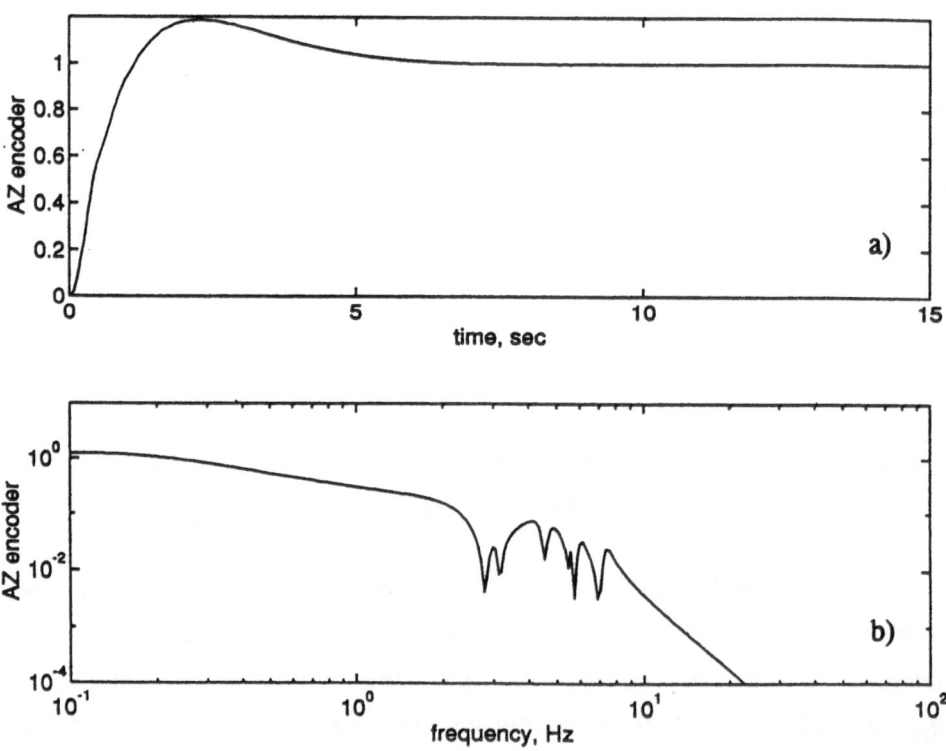

Figure 8.17: Antenna (a) step response, and (b) transfer function, for unit PI weights, and "flexible" weights 10^{-7}

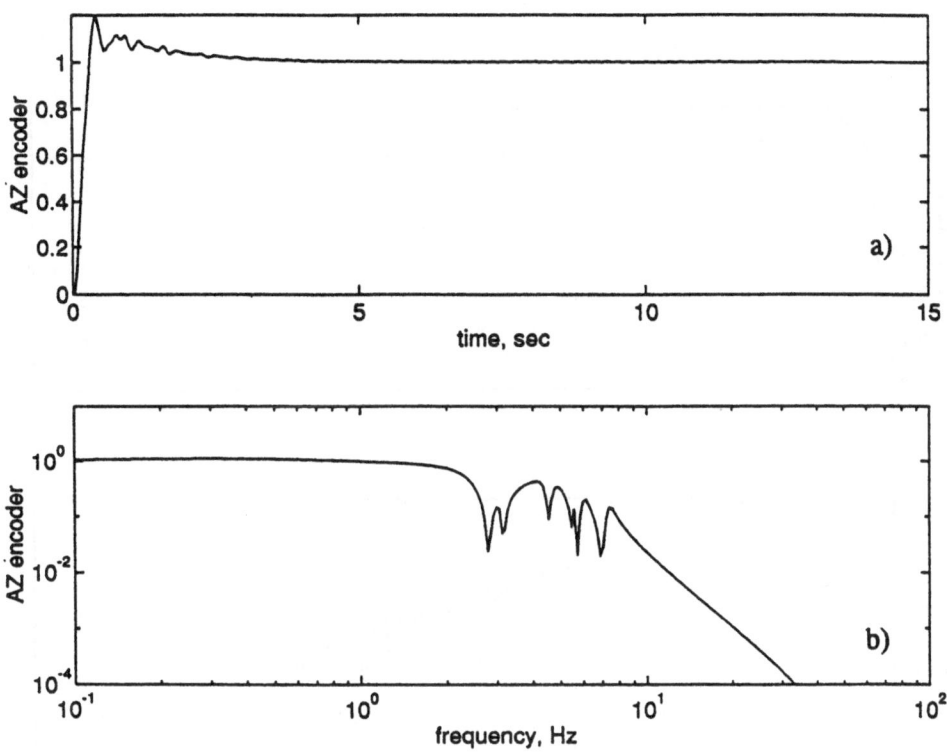

Figure 8.18: Antenna (a) step response, and (b) transfer function, for PI weights *100* and *70*, and "flexible" weights 10^{-7}

The reduced-order compensator is obtained through evaluation of controller reduction indices σ_i. The plot of σ_i is shown in Fig.8.20. Reducing the order of the estimator to 12 states (the first four are tracking states, the next six are flexible states, and the last two states are non-flexible components of the rate-loop model) yields a stable and accurate closed-loop system. The reduced-order compensator is compared with the full-order compensator in the step response plots in Fig.8.21a, and the transfer function plots in Fig.8.21b, showing satisfactory approximation.

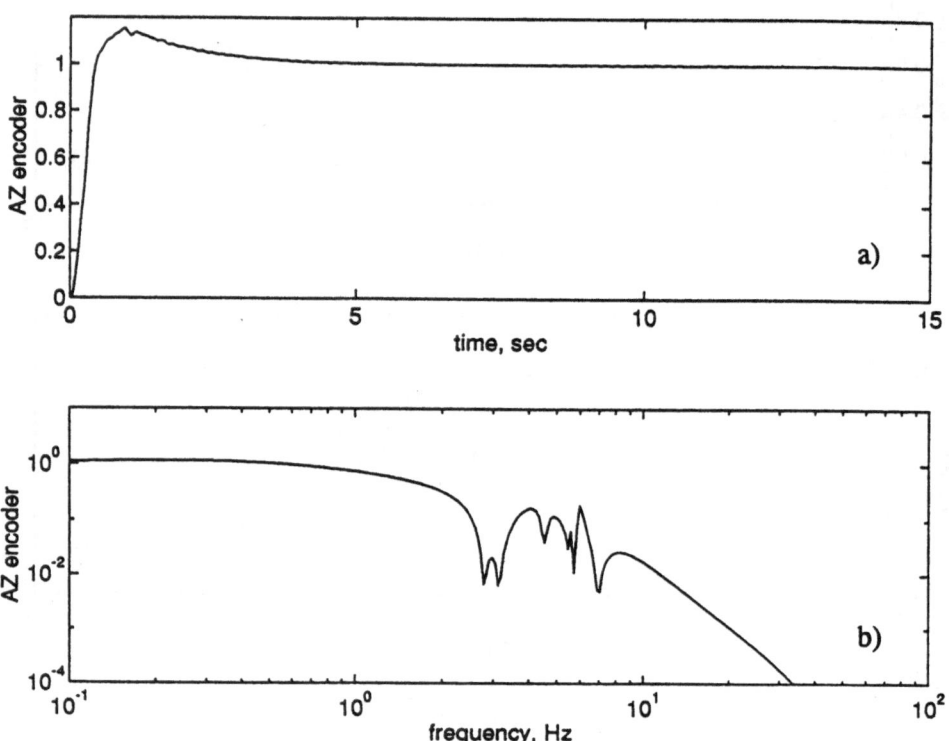

Figure 8.19: Antenna (a) step response, and (b) transfer function, for PI weights *100* and *70*, and "flexible" weights $q_{f1}=...=q_{f6}=10^{-6}$, $q_{f7}=q_{f8}=10^{-7}$, $q_{f9}=q_{f10}=10^{-5}$

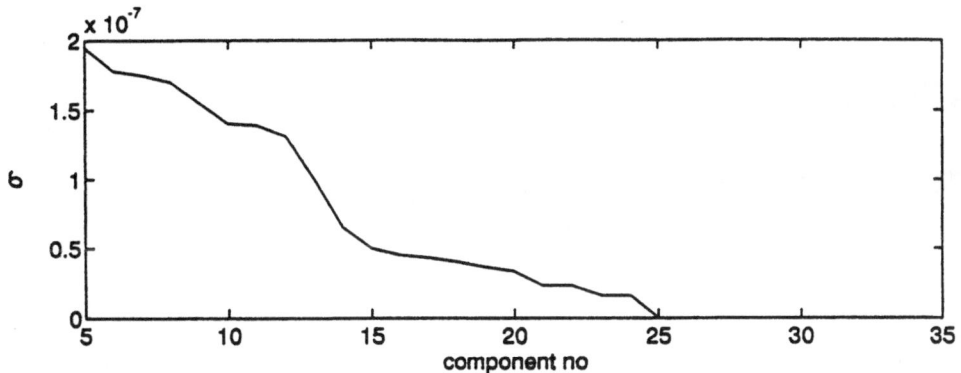

Figure 8.20: Reduction indices for the Deep Space Network antenna

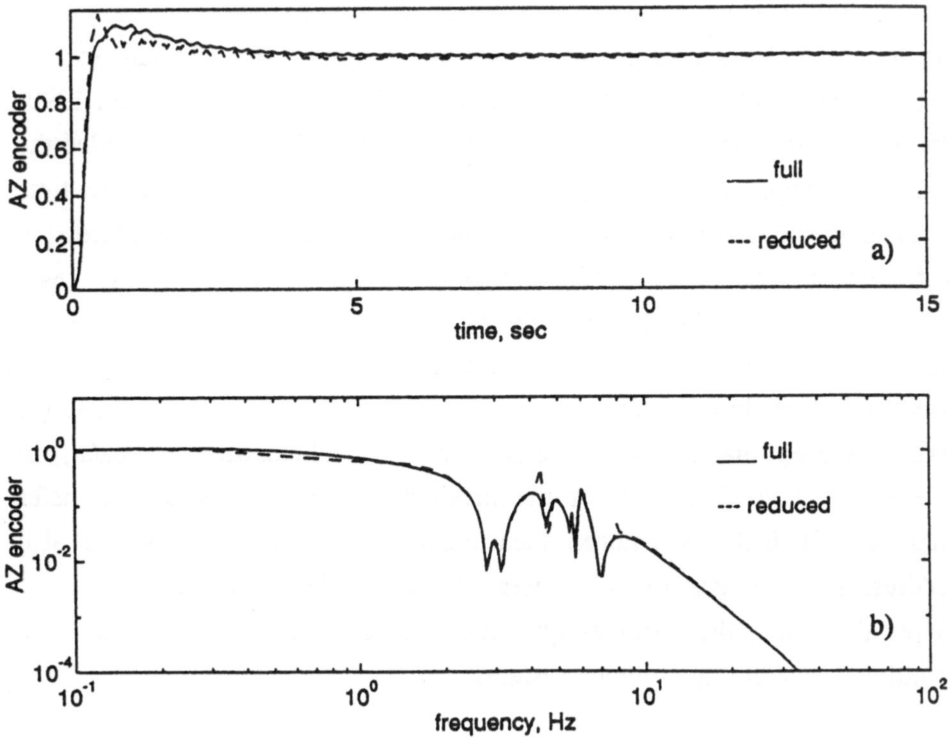

Figure 8.21: Antenna (a) step responses and (b) magnitudes of transfer function of the full- and reduced-order LQG controller

Chapter 9

Balanced H$_\infty$ Controllers

In 1940s and 1950s, at the beginning of the development of the control theory the frequency domain methods dominated the control system analysis and design. Later, in 1960s and 1970s the time domain (or the state-space) methods were extensively developed. And in 1980s and 1990s the H_∞ methods were introduced and developed. This method expanded the frequency approach, and also unified both the frequency and the time domain methods. Many books and papers have been published addressing different aspects of the H_∞ controller design, and Refs.[25], [84], [78], [11], and [108] explain the basic issues of the method. The H_∞ method addresses wide range of the control problems, combining the frequency and time domain approaches. The design is an optimal one (or rather sub-optimal, if one prefers strictly mathematical approach) in the sense of minimization of the maximum of the closed-loop transfer function. It includes colored measurement and process noise. It also addresses the issues of robustness due to model uncertainties, and is applicable to the single-input-single-output as well as to the multiple-input-multiple-output systems.

In this chapter the H_∞ controller design for flexible structures is presented. A balanced approach to the H_∞ controller design is chosen, since this approach allows for the determination of a reduced-order H_∞ controller which is stable, and which performance close to the full-order controller.

9.1 The H_∞ Norm of a Flexible Structure

Before discussing the H_∞ controllers, the determination of the H_∞ norm of a flexible structure is reviewed. Let (A,B,C) be a state-space triple of a flexible structure, and $G(\omega)=C(j\omega I-A)^{-1}B$ be its transfer function. Restricting for a moment our consideration to a single-input-single-output system, we obtain the H_∞ norm of G as the maximal value of the magnitude of the transfer function over all frequencies ω. In a single-input (u), single-output (z) case the H_∞ norm can also be interpreted in terms of energy, see Ref.[71], since

$$\|G\|_\infty^2 = \sup \left(\int_{-\infty}^{\infty} z^2(t)dt \right), \quad \text{provided that} \quad \int_{-\infty}^{\infty} u^2(t)dt \leq 1 \tag{9.1}$$

therefore

$$\|G\|_\infty^2 \geq \frac{\displaystyle\int_{-\infty}^{\infty} z^2(t)dt}{\displaystyle\int_{-\infty}^{\infty} u^2(t)dt} = \frac{output\ energy}{input\ energy} \tag{9.2}$$

The H_∞ norm contrasts with the H_2 norm, defined for a single-input-single-output system as $\|G\|_2^2 = \int_{-\infty}^{\infty} |G(j\omega)|^2 d\omega$. Hence, $\|G\|_2$ is proportional to the total energy in the impulse response.

For multiple-input and/or multiple-output systems, the H_∞ norm is defined as follows

$$\|G\|_\infty = \sup_{\omega} \sigma_{max}(G(j\omega)) \tag{9.3}$$

The H_∞ norm can be also evaluated from the Riccati equation, as shown by Doyle et al. [24], and Boyd and Barratt [11]. It is the smallest positive parameter ρ such that the solution of the following Riccati equation is positive definite

$$A^TS+SA+\rho^{-2}SBB^TS+C^TC=0. \tag{9.4}$$

This approach is computationally effective, however, the specific properties of a flexible structure allow for a simple determination of its H_∞ norm. We will show that from the definition (9.4) and the properties of flexible structures it follows that the H_∞ norm of a flexible structure is

$$\|G\|_\infty \cong 2\gamma_1^2 \tag{9.5}$$

where γ_1 is the largest Hankel singular value of the system.

Proof follows from the properties (3.60) and (3.61). In this case the solution S of the Riccati equation (9.4) is diagonally dominant, $S \cong diag(s_1, s_1, \ldots, s_n, s_n)$, where by inspection one can find s_i as a solution of the following equation

$$s_i(A_i+A_i^T)+s_i^2\rho_i^2B_iB_i^T+C_i^TC_i\cong0, \quad i=1,\ldots,n_2 \tag{9.6}$$

and A_i is given in Eq.(2.13b), B_i is a two-row block of B corresponding to block A_i of A, and C_i is a two-column block of C corresponding to block A_i of A. Introducing Eq.(3.61) to Eq.(9.6) one obtains

$$s_i^2(A_i+A_i^T)-s_i\gamma_i^2\rho_i^2(A_i+A_i^T)-\gamma_i^2(A_i+A_i^T)\cong0 \tag{9.7}$$

or

$$s_i^2 - \frac{\rho_i^2}{\gamma_i^2}s_i + \rho_i^2 \cong 0 \tag{9.8}$$

198

with two solutions $s_i^{(1)}$ and $s_i^{(2)}$

$$s_i^{(1)} \cong \frac{\rho_i^2(1-\beta_i)}{2\gamma_i^2}, \quad s_i^{(2)} \cong \frac{\rho_i^2(1+\beta_i)}{2\gamma_i^2}, \quad \beta_i = \sqrt{1 - \frac{4\gamma_i^4}{\rho_i^2}} \qquad (9.9)$$

For $\rho_i = 2\gamma_i^2$ one obtains $s_i^{(1)} = s_i^{(2)} = 2\gamma_i^2$. Moreover, $\rho_i = 2\gamma_i^2$ is the smallest value of ρ_i for which a positive solution s_i exists. It is indicated by the plots of $s_i^{(1)}$ (solid line) and $s_i^{(2)}$ (dashed line) vs ρ_i in Fig.9.1 for $\gamma_i = 1$. To obtain S positive definite, all s_i must be positive. Thus the the largest ρ from the set $\{\rho_1, \rho_2, \ldots, \rho_n\}$ is the smallest one for which S is positive definite, therefore $\|G\|_\infty = \max_i \rho_i \cong 2\gamma_i^2$.

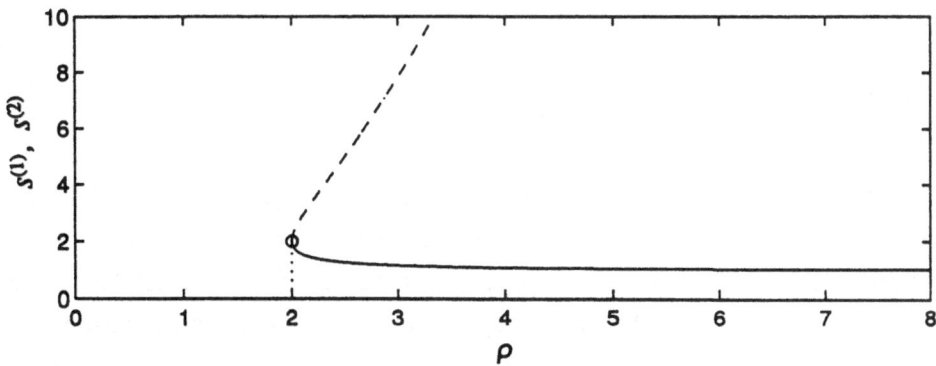

Figure 9.1: Solution s_i vs ρ_i for $\gamma_i = 1$

Example 9.1 A truss structure as in Fig.2.2 is investigated. Vertical control forces are applied at nodes $n7$ and $n8$, and the output rates are measured in the vertical direction at nodes $n3$ and $n4$. The system has 26 states, two inputs, and two outputs. Its H_∞ norm is computed through iterations of Riccati equation (9.4), obtaining $\|G\|_\infty = 1.340852$. It took 24 iterations to obtain the required accuracy $\varepsilon = 10^{-6}$, as shown in Fig.9.2. The H_∞ norm is also obtained from (9.5), and in this case $\|G\|_\infty = 1.340850$ with the same accuracy $\varepsilon = 10^{-6}$.

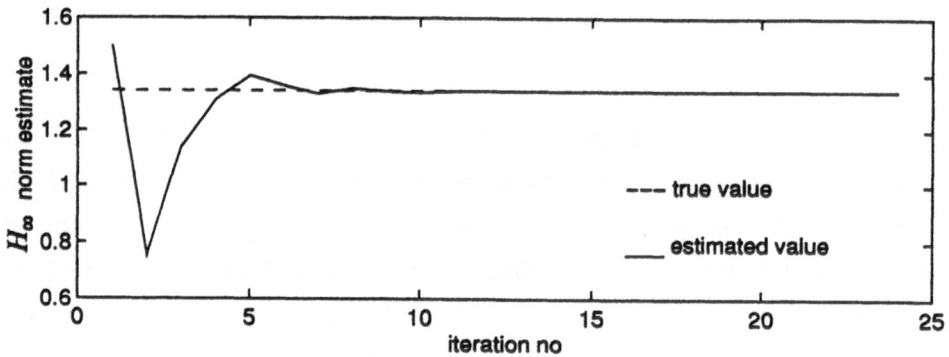

Figure 9.2: Estimated and accurate H_∞ norm

9.2 Definition and Properties

The closed-loop system architecture is shown in Fig.9.3, where G is the transfer function of a plant, K is the transfer function of a controller, w is the exogenous input (commands, disturbances), u is the actuator input, z is the regulated output (at which performance is evaluated), and y is the sensed output. The H_∞ control problem consists of determining K such that the H_∞ norm of the closed-loop transfer function from w to z, G_{zw}, is minimized over all realizable controllers K, that is

$$\text{find realizable } K \text{ such that } \|G_{zw}(K)\|_\infty \text{ is minimal} \qquad (9.10)$$

One of the ways to obtain the optimal gains of an H_∞ controller is to solve the constrained Riccati equations, see Ref.[24], as presented later.

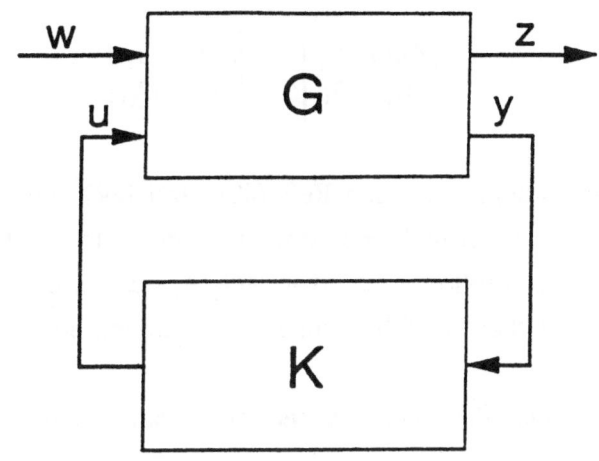

Figure 9.3: Closed-loop system configuration

9.2.1 Standard H_∞ Controller

Consider a representation of a closed-loop H_∞ system as in Fig.9.3, with the plant transfer function $G(s)$, and the controller transfer function $K(s)$, such that

$$\begin{pmatrix} z(s) \\ y(s) \end{pmatrix} = G(s) \begin{bmatrix} w(s) \\ u(s) \end{bmatrix}, \quad u(s) = K(s)y(s) \tag{9.11}$$

and u, w are control and exogenous inputs, and y, z are measured and controlled outputs, respectively. In the related state-space equations

$$\dot{x} = Ax + B_1 w + B_2 u, \quad z = C_1 x + D_{12} u, \quad y = C_2 x + D_{21} w \tag{9.12}$$

(A,B_2) is stabilizable and (A,C_2) is detectable, the conditions

$$D_{12}^T[C_1 \ D_{12}] = [0 \ I], \quad D_{21}[B_1^T \ D_{21}^T] = [0 \ I] \tag{9.13a}$$

are satisfied, and the matrices

$$\begin{bmatrix} A-j\omega I & B_2 \\ C_1 & D_{12} \end{bmatrix}, \quad \begin{bmatrix} A-j\omega I & B_1 \\ C_2 & D_{21} \end{bmatrix} \tag{9.13b}$$

have full column rank, see Refs.[49], and [24]. Let G_{zw} be the transfer function of the closed-loop system from w to z, then there exists an admissible controller such that $\|G_{zw}\|_\infty < \rho$, where ρ is the smallest number such that the following four conditions hold:

1. $S_{\infty c} \geq 0$ solves the following central Riccati equation (HCARE)

$$S_{\infty c} A + A^T S_{\infty c} + C_1^T C_1 - S_{\infty c} (B_2 B_2^T - \rho^{-2} B_1 B_1^T) S_{\infty c} = 0 \tag{9.14a}$$

2. $S_{\infty e} \geq 0$ solves the following central Riccati equation (HFARE)

$$S_{\infty e} A^T + A S_{\infty e} + B_1 B_1^T - S_{\infty e} (C_2^T C_2 - \rho^{-2} C_1^T C_1) S_{\infty e} = 0 \tag{9.14b}$$

3. $\qquad\qquad\qquad \lambda_{max}(S_{\infty c} S_{\infty e}) < \rho^2 \tag{9.14c}$

where $\lambda_{max}(X)$ is the largest eigenvalue of X, and

4. The Hamiltonian matrices

$$\begin{bmatrix} A & \rho^{-2} B_1 B_1^T - B_2 B_2^T \\ -C_1^T C_1 & -A^T \end{bmatrix}, \quad \begin{bmatrix} A^T & \rho^{-2} C_1^T C_1 - C_2^T C_2 \\ -B_1 B_1^T & -A \end{bmatrix} \tag{9.14d}$$

do not have eigenvalues on the $j\omega$ axis.

With the above conditions satisfied, the controller realization $(A_\infty, B_\infty, C_\infty)$, from input y to output u, is given as

202

$$A_\infty = A + \rho^{-2} B_1 B_1^T S_{\infty c} + B_2 k_c + k_e C_2, \qquad B_\infty = -k_e, \qquad C_\infty = k_c, \qquad (9.15a)$$

where

$$k_c = -B_2^T S_c, \qquad k_e = -S_o S_{\infty e} C_2^T, \qquad S_o = (I - \rho^{-2} S_{\infty e} S_{\infty c})^{-1} \qquad (9.15b)$$

The gain k_c is called the control gain, while the gain k_e is the filter (estimator) gain. Note that the form of the solution is similar to the LQG solution. However the LQG gains are determined independently, while the H_∞ gains are coupled through the inequality (9,16c), and the component S_o. The size of the controller is equal to the size of the plant.

9.2.2 Balanced H_∞ Controller

A balanced H_∞ controller was investigated by Mustafa and Glover (see Refs.[90], and [91]) for a special case of collocated control and exogenous inputs, and collocated measured and controlled outputs. This case, is in fact a modified LQG controller. Indeed, for $B_1 = B_2$ and $C_1 = C_2$, the HCARE and HFARE equations become the CARE and FARE equations, respectively, with the scaling factor $(1-\rho^{-2})$ at the second-order term. Here, following Ref.[40] a general case of the central H_∞ controller is considered.

An H_∞ controller is balanced if the related HCARE and HFARE solutions are equal and diagonal, i.e., if

$$S_{\infty c} = S_{\infty e} = M_\infty, \qquad M_\infty = diag(\mu_{\infty 1}, \mu_{\infty 2}, \dots, \mu_{\infty N}), \qquad (9.16)$$

$\mu_{\infty 1} \geq \mu_{\infty 2} \geq \dots \geq \mu_{\infty N} > 0$, where $\mu_{\infty i}$ is the ith H_∞ characteristic (or singular) value.

The H_∞ balanced representation exists, and the transformation R_∞

to this representation is determined as follows. Let

$$P_{\infty c} = S_{\infty c}^{1/2}, \quad P_{\infty e} = S_{\infty e}^{1/2} \tag{9.17a}$$

denote $N_\infty = P_{\infty c} P_{\infty e}$, and let

$$N_\infty = V_\infty M_\infty U_\infty^T \tag{9.17b}$$

be the singular value decomposition of N. For the transformation R_∞

$$R_\infty = P_{\infty e} U_\infty M_\infty^{-1/2} = P_{\infty c}^{-1} V_\infty M_\infty^{1/2} \tag{9.18}$$

of the state x such that $\bar{x} = R_\infty x$, the representation $(R_\infty^{-1} A R_\infty, \ R_\infty^{-1} B_1,$ $R_\infty^{-1} B_2, \ C_1 R_\infty, \ C_2 R_\infty)$ is H_∞ balanced.

In order to prove it, note that the solutions of HCARE and HFARE in new coordinates are $\bar{S}_{\infty c} = R_\infty^T S_{\infty c} R_\infty, \ \bar{S}_{\infty e} = R_\infty^{-1} S_{\infty e} R_\infty^{-T}$. Introducing R_∞ as in Eq.(9.18) gives the balanced HCARE and HFARE solutions.

The Matlab function *bal_H_inf* which transforms a representation (A, B_1, B_2, C_1, C_2) to the H_∞ balanced representation $(A_b, B_{b1}, B_{b2}, C_{b1}, C_{b2})$ is given in the Appendix B5.

For the H_∞ balanced solution the condition in Eq.(9.14c) simplifies to

$$\mu_{\infty 1} < \rho, \text{ and } \mu_{\infty n} > 0 \tag{9.19}$$

Denote $X_1 > X_2$ $(X_1 \geq X_2)$ for $X_1 - X_2$ positive definite (positive semi-definite). The relationship between H_∞ characteristic values and open-loop (Hankel) singular values is established. First, for the asymptotically stable A, for $V > 0$, for two Riccati equations:

$$A^T S_i + S_i A - S_i W_i S_i + V = 0, \quad i = 1, 2 \tag{9.20}$$

and for $W_2 \geq W_1 \geq 0$, one obtains

$$S_1 \geq S_2 \geq 0. \tag{9.21}$$

see Ref.[21].

Let Γ_1 be a matrix of Hankel singular values of the representation (A, B_1, C_1), and M_∞ be a matrix of H_∞ characteristic values defined in Eq.(9.16). Then for asymptotically stable A, and for $B_2 B_2^T - \rho^{-2} B_1 B_1^T \geq 0$, $C_2^T C_2 - \rho^{-2} C_1^T C_1 \geq 0$, one obtains

$$M_\infty \leq \Gamma_1^2, \quad \text{or} \quad \mu_{\infty i} \leq \gamma_{1i}^2, \quad i=1,\dots,N \tag{9.22}$$

In order to prove it note that Eq.(9.22) is a consequence of the property as given in Eq.(9.21), which is applied to Eq.(9.14a), and to the Lyapunov equation $W_o A + A^T W_o + C_1^T C_1 = 0$. It is also a consequence of the property (9.21) applied to Eq.(9.14b), and to the Lyapunov equation $W_c A^T + A W_c + B_1 B_1^T = 0$. In this way one obtains $W_{c1} \geq S_{\infty c}$ and $W_{o1} \geq S_{\infty c}$. From the latter inequalities it follows that $\lambda_i(W_{c1}) \geq \lambda_i(S_{\infty c})$ and $\lambda_i(W_{o1}) \geq \lambda_i(S_{\infty c})$ (see Ref.[54], p.471), thus $\lambda_i(W_{c1} W_{o1}) \geq \lambda_i(S_{\infty c} S_{\infty c})$, or $M_\infty \leq \Gamma_1^2$.

9.2.3 H_2 Controller

An H_2 system is a special case of the H_∞ system, c.f. Ref.[11]. It has representation similar to the H_∞ system as in Eq.(9.12), and its matrices A, B_1, B_2, C_1, C_2, D_{12}, D_{21} are defined in the following. It consists of state x, control input u, measured output y, exogenous input $w^T = [v_u^T \ v_y^T]$, and regulated variable $z = C_1 x + D_{12} u$, where v_u and v_y are process and measurement noise, respectively. The noises v_u and v_y are uncorrelated, and have constant power spectral density matrices V_u and V_y, respectively. For positive semi-definite matrix V_u, the matrix B_1 has the following form:

$$B_1 = [V_u^{1/2} \quad 0] \tag{9.23}$$

The task is to determine the controller gain (k_c) and estimator gain (k_e) such that the performance index J as in Eq.(8.1) is minimal, where R is a positive definite input weight matrix, and Q a positive semi-definite state weight matrix. Matrix C_1 is defined through the weight Q

$$C_1 = \begin{bmatrix} 0 \\ Q^{1/2} \end{bmatrix} \tag{9.24}$$

and, without loss of generality, assume $R=I$ and $V_y=I$, obtaining

$$D_{12} = \begin{bmatrix} I \\ 0 \end{bmatrix}, \quad D_{21} = [0 \quad I] \tag{9.25}$$

The minimum of J is achieved for the feedback where the gain matrices $(k_c$ and $k_e)$

$$k_c = -B_2^T S_{2c}, \quad k_e = -S_{2e} C_2^T \tag{9.26}$$

where S_{2c} and S_{2e} are solutions of the controller Riccati equation (8.3a) (CARE) and the estimator Riccati equation (8.3b) (FARE), respectively, which in this case are as follows

$$S_{2c}A + A^T S_{2c} + C_1^T C_1 - S_{2c} B_2 B_2^T S_{2c} = 0 \tag{9.27a}$$

$$S_{2e}A^T + A S_{2e} + B_1 B_1^T - S_{2e} C_2^T C_2 S_{2e} = 0 \tag{9.27b}$$

Note by comparing Eqs.(9.14) and (9.27) that for $\rho^{-1}=0$ the H_∞ solution becomes the H_2 solution.

9.2.4 Balanced H_2 Controller

A balanced H_2 (or LQG) controller was discussed in Chapter 8, and here its properties are compared with properties of the H_∞ controller. An H_2 controller is balanced if the related CARE and FARE solutions are equal and diagonal. The relationship between H_∞ and H_2 characteristic values is derived as follows.

Let $\beta = inf\{\rho: M_\infty(\rho) \geq 0\}$. Then on the segment $(\beta, +\infty)$ all H_∞ characteristic values, $\mu_{\infty i}$ $i=1,...,n$, are smooth nonincreasing functions of ρ, and the maximal characteristic value $\mu_{\infty 1}$ is a nonincreasing convex function of ρ, see Ref.[72]. As a consequence, for $\rho \to \infty$, one obtains $M_\infty \to M_2$. However, $\mu_{\infty i}$ are increasing functions of ρ, and $\mu_{\infty i} \to \mu_{2i}$ as $\rho \to \infty$, thus $\mu_{2i} \leq \mu_{\infty i}$, or

$$M_2 \leq M_\infty, \quad \text{or} \quad \mu_{2i} \leq \mu_{\infty i}, \quad i=1,...,N \tag{9.28}$$

The H_2 characteristic values and Hankel singular values are also related, as follows from Eq.(9.22) for $\rho \to \infty$

$$M_2 \leq \Gamma_1^2, \quad \text{or} \quad \mu_{2i} \leq \gamma_{1i}^2, \quad i=1,...,N \tag{9.29}$$

9.3 Balanced H_∞ Controller for Flexible Structures

In a preliminary move to solve the H_∞ control problem for flexible structures consider two systems: (A,B_1,C_1) and (A,B_2,C_2). The first represents the plant as in Fig.9.3 with input w and output z, the second represents the same plant with input u and output y. Thus, denote (A,B_k,C_k) and $k=1,2$, the state-space representations of stable, controllable, and observable open-loop systems, where A is $n \times n$, B_k is $n \times p_k$, and C_k is $q_k \times n$. Their controllability and observability grammians W_{ck} and W_{ok} are positive-definite and satisfy the Lyapunov equations

$$AW_{ck}+W_{ck}A^T+B_kB_k^T=0, \quad A^TW_{ok}+W_{ok}A+C_k^TC_k=0 \qquad (9.30)$$

for $k=1,2$.

The system representations are balanced in the sense of Moore i.e., their controllability and observability grammians are diagonal and equal

$$W_{ck}=W_{ok}=\Gamma_k^2, \quad \Gamma_k=diag(\gamma_{k1},...,\gamma_{kN}), \quad k=1,2 \qquad (9.31)$$

and $\gamma_{kj}>0$ is the jth Hankel singular value of the kth system.

Consider the modal representations of the two systems in form 2 (A_m,B_{m1},C_{m1}) and (A_m,B_{m2},C_{m2}), where matrix A_m is block-diagonal, with 2×2 blocks on the main diagonal

$$A_m=diag(A_i), \quad A_i=\begin{bmatrix} -\zeta_i\omega_i & -\omega_i \\ \omega_i & -\zeta_i\omega_i \end{bmatrix}, \quad i=1,...,n \qquad (9.32)$$

where ω_i is the ith natural frequency of the structure, and ζ_i is the ith modal damping. The matrices B_{m1}, B_{m2}, C_{m1}, and C_{m2} are not unique - they depend on normalization of the modal matrix. The modal and the open-loop balanced coordinates are almost identical, and they required re-scaling only, as discussed in Chapter 3. In fact, denote the first balanced representation $(A_{b1}^{(1)}B_{b1}^{(1)}C_{b1}^{(1)})$ and the second balanced representation $(A_{b2}^{(2)}B_{b2}^{(2)}C_{b2}^{(2)})$. Denote also the first system in the second balanced representation as $(A_{b1}^{(2)}B_{b1}^{(2)}C_{b1}^{(2)})$, and the second system in the first balanced representation as $(A_{b2}^{(1)}B_{b2}^{(1)}C_{b2}^{(1)})$. The transformation from the modal representation (A_m,B_{mk},C_{mk}) to the balanced representation $(A_{bk}^{(k)}B_{bk}^{(k)}C_{bk}^{(k)})$ is denoted R_{mk}, $k=1,2$

$$(A_{bk}^{(k)}B_{bk}^{(k)}C_{bk}^{(k)}) \cong (A_m,R_{mk}^{-1}B_{mk},C_{mk}R_{mk}) \qquad (9.33a)$$

and is diagonally dominant. In order to measure the dominance of the

208

diagonal terms of R_{mk} define

$$\varepsilon_{mki} = \frac{r_{mki}}{\| r_{mki} \|}, \quad i=1,\ldots,n \qquad (9.33b)$$

where r_{mki} is the ith colon of R_{mk}. For a diagonally dominant R_{mk} one obtains $\varepsilon_{mki} \cong e_i$, where e_i is the unit vector (all but one zero components, the nonzero component equal to 1). Its diagonal entries depend on the scaling of the modal matrix Φ. The transformation from modal to the balanced coordinates are illustrated in Fig.9.4.

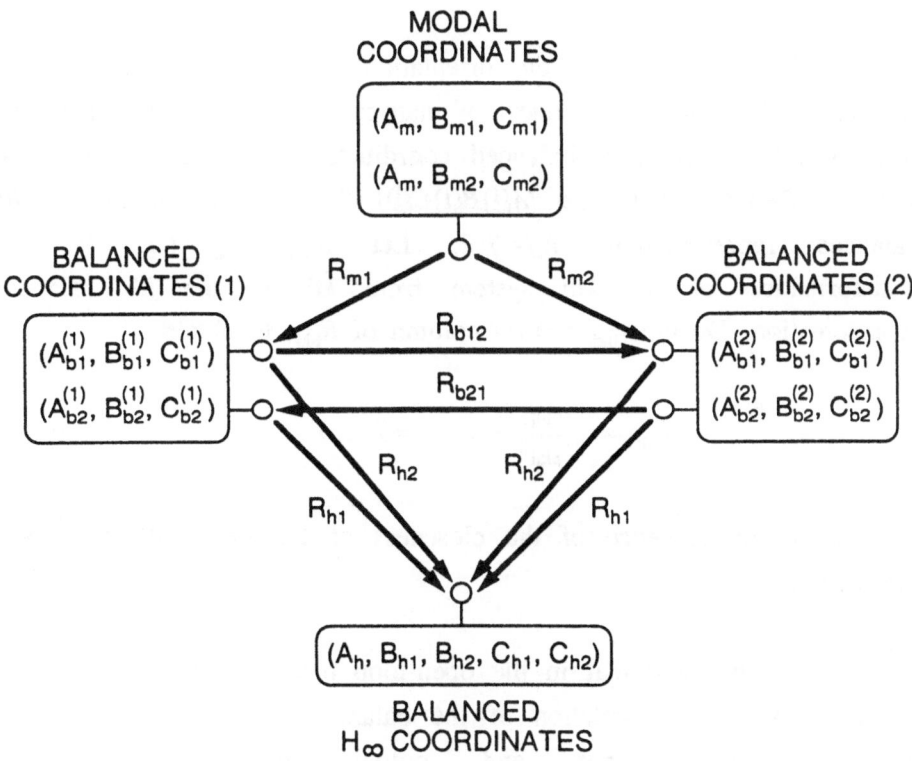

Figure 9.4: Coordinate transformations for flexible structures

For the state-space representations of a flexible structure (A,B_k,C_k), $k=1,2$ the controllability and observability grammians W_{ck} and W_{ok} are positive definite and satisfy the Lyapunov equations (9.30). For a flexible structure with n_2 components (or $n=2n_2$ states), the balanced grammian has the following form

$$\Gamma_k \cong diag(\gamma_{ki}I_2), \quad k=1,2, \quad i=1,\dots,n \tag{9.34}$$

Matrix A is almost block-diagonal, with dominant 2×2 blocks on the main diagonal, $A \cong diag(A_i)$, $i=1,\dots,n$, where A_i is given in Eq.(9.32). Introducing Eqs.(9.34) and A as above to Eq.(9.30) gives

$$B_{ki}B_{ki}^T \cong C_{ki}^T C_{ki} \cong -\gamma_{ki}^2(A_i+A_i^T)=2\zeta_i\omega_i\gamma_{ki}^2 I_2. \tag{9.35}$$

For flexible structures the orientation of the open-loop balanced coordinates is almost independent of matrices B_k and C_k, and the matrix A is almost invariant in balanced coordinates. This can be stated as follows. Denote as before $(A_{bk}^{(j)}B_{bk}^{(j)}C_{bk}^{(j)})$ the kth system in the jth balanced representation, $k,j=1,2$. Let R_{bkj}, $k,j=1,2$, be the transformation of the kth system from kth to the jth balanced representation. Denote r_{bkji} the ith column of R_{bkj}, to obtain

$$\varepsilon_{bkji} = \frac{r_{bkji}}{\|r_{bkji}\|} \cong e_i, \quad i=1,\dots n \tag{9.36}$$

as a direct consequence of the closeness of the balanced and modal representation.

The results show that in the open-loop balanced representation the matrix A and the orientation of the balanced coordinates are almost invariant under input and output locations. That is, $A_{b2}^{(2)} \cong A_{b2}^{(1)} \cong A_{b1}^{(2)} \cong A_{b1}^{(1)} \cong A_b$, and the transformations from the one to the another balanced representations are as follows

$$(A_{b1}^{(2)}B_{b1}^{(2)}C_{b1}^{(2)}) \cong (A_{b1}^{(1)}R_{b12}^{-1}B_{b1}^{(1)}C_{b1}^{(1)}R_{b12}) \tag{9.37a}$$

210

$$(A_{b2}^{(1)}B_{b2}^{(1)}C_{b2}^{(1)}) \cong (A_{b2}^{(2)}R_{b21}^{-1}B_{b2}^{(2)}C_{b2}^{(2)}R_{b21})$$ (9.37b)

The transformation matrix is diagonally dominant, i.e., requires only a re-scaling of the coordinates

$$R_{b12} \cong diag(r_1 I_2, \ r_2 I_2, \ ..., \ r_n I_2), \qquad r_i = \frac{\gamma_{1i}^2}{\gamma_{2i}^2}$$ (9.37c)

$$R_{b21} = R_b^{-1}|_2$$ (9.37d)

where γ_{1i}, γ_{2i} are the ith Hankel singular values of the first and the second system, respectively. The transformations between the open-loop balanced representations are shown in Fig.9.4.

The properties of flexible structures, specified above, are now extended to H_∞ balanced flexible structures. Denote a system quintuple in kth balanced coordinates as $(A_b, B_{b1}^{(k)}, B_{b2}^{(k)}, C_{b1}^{(k)}, C_{b2}^{(k)})$. Let R_{hk}, $k=1,2$ be now a transformation of the quintuple from the kth balanced representation to the H_∞ balanced representation, and let r_{hki} be the ith column of R_{hk}. Then, for the flexible structures one obtains

$$\varepsilon_{hki} = \frac{r_{hki}}{\|r_{hki}\|} \cong e_i \qquad k=1,2, \quad i=1,...,n$$ (9.38)

and the solutions of HCARE and HFARE in kth open-loop balanced coordinates ($k=1,2$) are diagonally dominant

$$S_{\infty ck} \cong diag(s_{\infty cki} I_2), \quad S_{\infty ek} \cong diag(s_{\infty eki} I_2), \quad i=1,2,...,n$$ (9.39)

In order to prove it, note that the diagonally dominant solutions of the Riccati equations (9.14a,b), follow from the properties of flexible structures, Eqs.(9.34) and (9.35). Thus the transformation matrices R_{hk} from open- to closed-loop balanced representation are diagonally dominant.

This shows also that the orientation of the H_∞ balanced coordinates is almost identical with the orientation of the open-loop balanced coordinates, and the open- and closed-loop balanced representations are related as follows:

$$(A_h, B_{h1}, B_{h2}, C_{h1}, C_{h2}) \cong (A_b, R_{hk}^{-1}B_1^{(k)}, R_{hk}^{-1}B_2^{(k)}, C_1^{(k)}R_{hk}, C_2^{(k)}R_{hk}) \qquad (9.40a)$$

where the transformation R_{hk} is as follows:

$$R_{hk} \cong diag(r_{hk1}I_2, \ldots, r_{hkn}I_2), \qquad r_{hki} = \left(\frac{S_{\infty eki}}{S_{\infty cki}}\right)^{1/4} \qquad (9.40b)$$

$k=1,2$, and the HCARE, HFARE solution M_∞ is diagonally dominant in the open-loop balanced coordinates

$$M_\infty \cong diag(\mu_{\infty i}I_2), \qquad \mu_{\infty i} = \sqrt{S_{\infty cki}S_{\infty eki}}, \quad i=1,\ldots,n \qquad (9.40c)$$

It is easily proven by noting that the solutions of HCARE and HFARE are $S_{\infty ckh} = R_{hk}^T S_{\infty ck} R_{hk}$, $S_{\infty ekh} = R_{hk}^{-1} S_{\infty ek} R_{hk}^{-T}$, and introducing R_{hk} as in Eq.(9.40b), one obtains a balanced solution as in Eq.(9.40c). Note that values of $S_{\infty cki}$ and $S_{\infty eki}$ depend on choice of coordinates, but their product does not. The transformations between the open-loop balanced representations and the H_∞ balanced representation is shown in Fig.9.4.

Note that for flexible structures, using Eqs.(9.34), (9.35) and (9.39), the balanced Riccati equations (9.14) can be written as follows

$$\kappa_i\mu_{\infty i}^2 + \mu_{\infty i} - \gamma_{1i}^2 \cong 0, \quad i=1,\ldots,n. \qquad (9.41a)$$

where

$$\kappa_i = \gamma_{2i}^2 - \frac{\gamma_{1i}^2}{\rho^2} \ . \tag{9.41b}$$

The solution of the ith equation

$$\mu_{\infty i} \cong \frac{-1 \pm \sqrt{1 + 4\gamma_{1i}^2 \kappa_i}}{2\kappa_i} \tag{9.42}$$

is real and positive for $\kappa_i > -0.25\gamma_{1i}^2$. From Eq.(9.41), one obtains $(2\kappa_i\mu_{\infty i} + 1)^2 \cong 1 + 4\gamma_{1i}^2\kappa_i$, or after simplifications $\kappa_i\mu_{\infty i}^2 + \mu_{\infty i} \cong \gamma_{1i}^2$. Thus

$$\mu_{\infty i} \leq \gamma_{1i}^2 \quad \text{for } \kappa_i \geq 0, \tag{9.43a}$$

$$\mu_{\infty i} > \gamma_{1i}^2 \quad \text{for } 0 > \kappa_i > -\frac{1}{4\gamma_{1i}^2} \tag{9.43b}$$

The above results can be specified for the H_2 systems by setting $\rho^{-1} = 0$. Thus for H_2 controller $\kappa_i \cong \gamma_{2i}^2$, and from Eq.(9.42), it follows that

$$\mu_{2i} \cong \frac{-1 + \sqrt{1 + 4\gamma_{1i}^2\gamma_{2i}^2}}{2\gamma_{2i}^2} \tag{9.44}$$

is the unique positive solution of the balanced H_2 Riccati equations. Thus μ_{2i} is the ith characteristic value of an H_2 system, as in Chapter 8. Also, from Eqs.(9.43) one obtains

$$\mu_{2i} \leq \mu_{\infty i} \leq \gamma_{1i}^2, \quad \text{and} \quad \mu_{21} \leq \rho \leq \gamma_{11}^2, \quad \text{for } \kappa_i \geq 0 \tag{9.45a}$$

$$\mu_{\infty i} > \gamma_{1i}^2, \quad \text{and} \quad \rho > \gamma_{11}^2, \quad \text{for } 0 > \kappa_i > -\frac{1}{4\gamma_{1i}^2} \tag{9.45b}$$

213

9.4 Frequency Weighting

Frequency weighting of the exogenous inputs and regulated outputs is a standard procedure in the H_∞ design to obtain the required closed-loop properties, see Ref.[25]. In this case a plant is augmented with the input and/or output shaping filters, just forming a new plant model, which consequently is used in the H_∞ design. For flexible structures, one can implement an approach similar to the LQG design, where the frequency weighting was equivalent to the replacement of the plant Hankel singular values with the Hankel singular values of the augmented plant. The replacement was achieved by simply rescaling matrices B_1 and C_1 as in Section 8.4. This approach is illustrated later with the H_∞ controller for a Deep Space Network antenna.

9.5 Closed-Loop System

The state-space equations of the open-loop system follow from Eq.(9.12)

$$\dot{x}=Ax+B_1w+B_2u, \quad z=C_1x+D_{12}u, \quad y=C_2x+D_{21}w, \quad u=k_c\hat{x} \qquad (9.46a)$$

and the state-space equations of the central H_∞ controller are as follows, see Refs.[49], and [24]:

$$\dot{\hat{x}}=(A+k_eC_2+\rho^{-2}B_1B_1^TS_{\infty c}+B_2k_c)\hat{x}-k_ey \qquad (9.46b)$$

where

$$k_c=-B_2^TS_{\infty c}, \quad k_e=-S_oS_{\infty e}C_2^T, \quad S_o=(I-\rho^{-2}S_{\infty e}S_{\infty c})^{-1} \qquad (9.47)$$

Defining a new state variable $x_o^T=[x^T \ \varepsilon^T]$, where $\varepsilon=x-\hat{x}$, one obtains the closed-loop state-space equations

$$\dot{x}_o=A_ox_o+B_ow, \quad z=C_ox_o \qquad (9.48a)$$

214

where

$$A_o = \begin{bmatrix} A+B_2k_c & -B_2k_c \\ -\rho^{-2}B_1B_1^T S_{\infty c} & A+k_eC_2+\rho^{-2}B_1B_1^T S_{\infty c} \end{bmatrix}, \qquad (9.48b)$$

$$B_o = \begin{bmatrix} B_1 \\ B_1-k_eD_{21} \end{bmatrix}, \qquad C_o = \begin{bmatrix} C_1+D_{12}k_c & -D_{12}k_c \end{bmatrix} \qquad (9.48c)$$

The block diagram of the closed-loop system is shown in Fig.9.5. Assuming $\rho^{-1}=0$ in (9.48), one obtains the H_2 system.

For the balanced H_∞ system one obtains from Eq.(9.47)

$$k_c = -B_2^T M_\infty, \quad k_e = -S_o M_\infty C_2^T, \quad S_o = (I-\rho^{-2}M_\infty^2)^{-1} \qquad (9.49a)$$

or, denoting $k_c=[k_{c1},...,k_{cn}]$, $k_e^T=[k_{e1}^T,...,k_{en}^T]$, $B_2^T=[B_{21}^T,...,B_{2n}^T]$, $C_2=[C_{21},...,C_{2n}]$, then

$$k_{ci}=-\mu_{\infty i}B_{2i}^T, \quad k_{ei}=-\mu_{\infty i}S_{oi}C_{2i}, \quad S_{oi}=\frac{\rho^2}{\rho^2-\mu_{\infty i}^2} \qquad (9.49b)$$

From Eq.(9.49b) it could be seen that the ith component of controller gain (k_{ci}) and estimator gain (k_{ei}) is weighted with the ith H_∞ characteristic value. The ith estimator gain is weighted additionally with s_{oi}. Since $s_{o1} \to \infty$ as $\rho \to \mu_{\infty 1}$ the first estimator gain, k_{e1}, can be arbitrarily large. But the increase of the gain k_{e1} will not greatly improve the system performance. In order to see it note that as ρ approaches $\mu_{\infty 1}$ the difference $\varepsilon_1=\rho^2-\mu_{\infty 1}^2$ approaches zero, and the performance, measured by $\|G_{zw}\|$, improves by negligible ε_1 (but the gain increase is of order ε_1^{-2}).

9.6 Tracking H_∞ Controller

The tracking control problem differs from the regulation problem because the controller performance depends not only on the plant parameters, but also on the tracking command profile (its rate, acceleration, etc.). Although the H_∞ control problem is flexible enough to formulate the tracking problem to meet requirements, it is useful to formulate it, such that some requirements are met by definition. One important requirement for tracking systems is to maintain zero steady-state error. This requirement can be satisfied by upgrading the plant with an integrator. This approach was already discussed in the LQG design procedure. An H_∞ tracking controller with an integral upgrade is presented in Fig.9.6, which is obviously different than the H_∞ controller in Fig.9.5. This design approach is similar to the LQG tracking controller design presented earlier.

9.7 Reduced-Order H_∞ Controller

The order of the central H_∞ controller is equal to the order of the plant, and often is too large for implementation. Order reduction is therefore an important design issue. The reduction of a generic H_∞ controller is not a straightforward task, however an H_∞ controller for flexible structures inherits special properties which can be used for controller reduction purposes.

9.7.1 Reduction Index

The introduction of the reduction index, or an indicator of importance of controller components, is necessary to make reasonable decisions concerning the controller order. The following reduction index is used here for the H_∞ controller reduction purposes

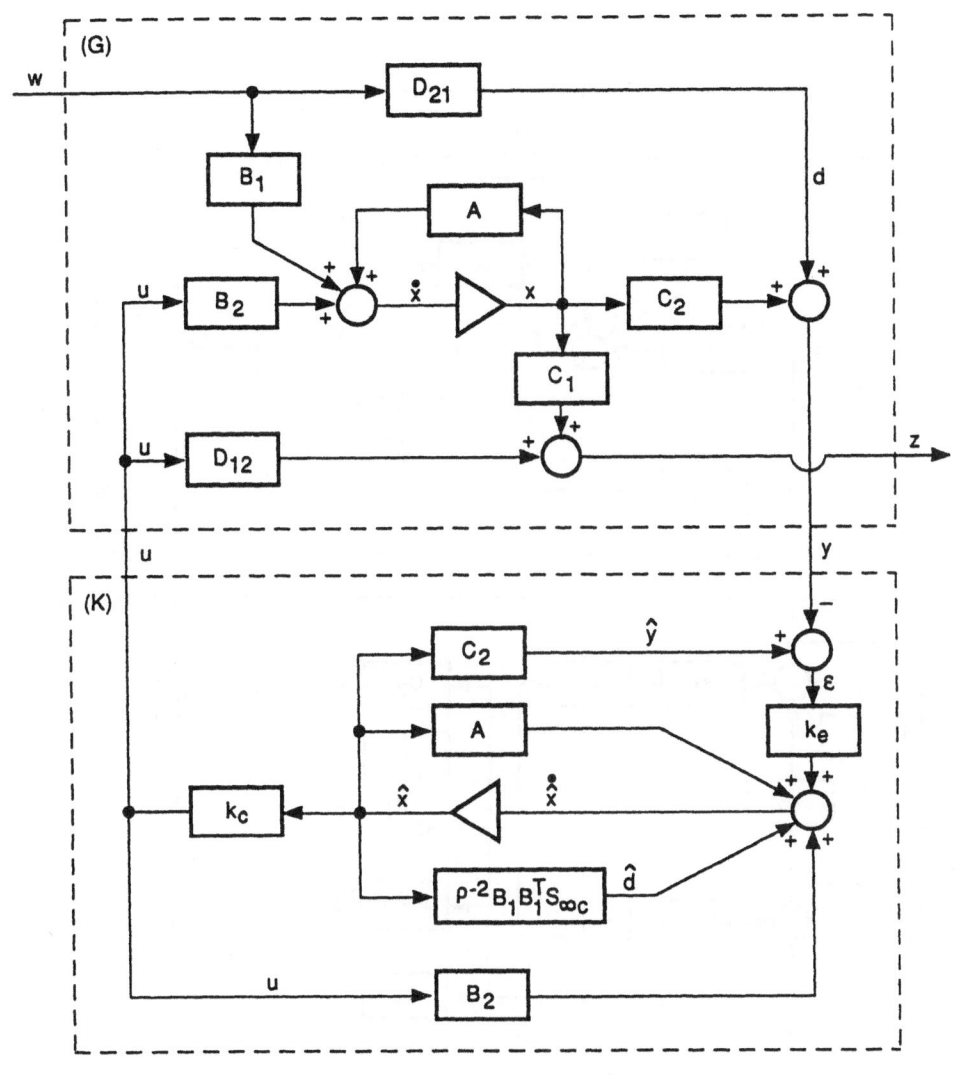

Figure 9.5: H_∞ standard closed-loop system

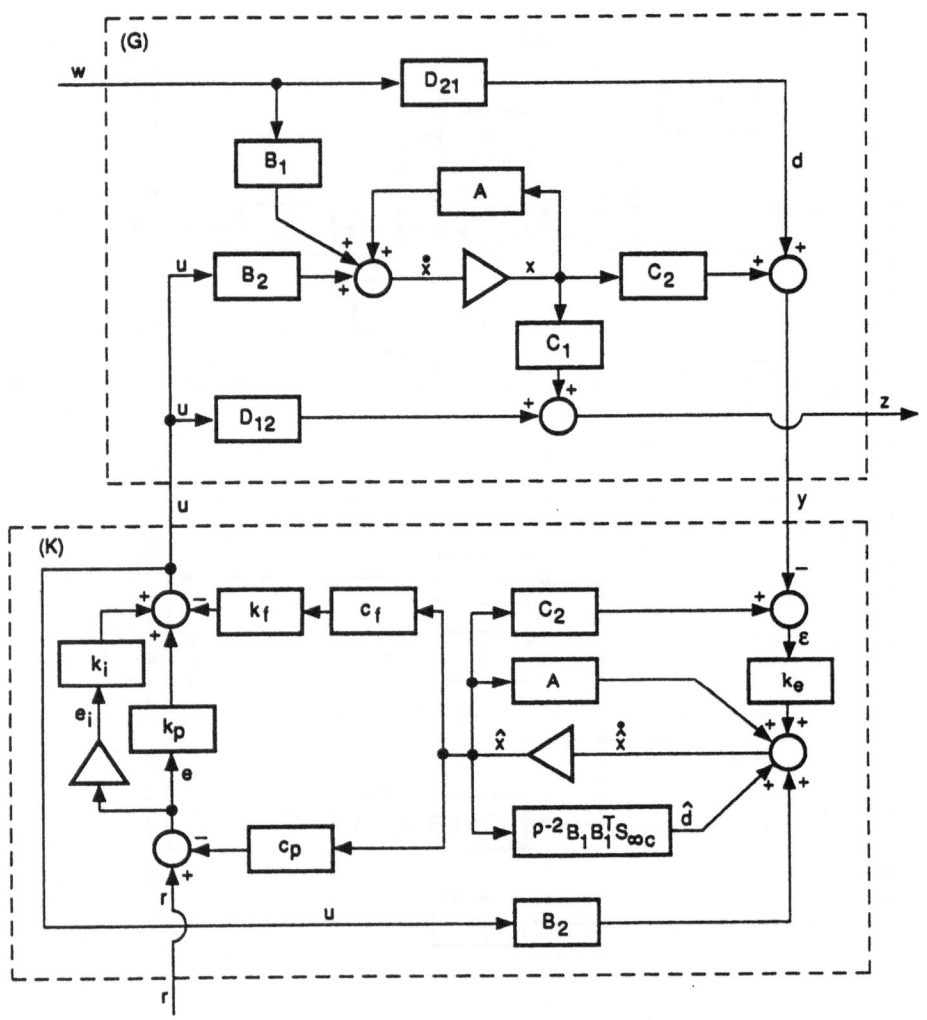

Figure 9.6: Tracking H_∞ standard closed-loop system

218

$$\sigma_{\infty i} = \gamma_{2i}^2 \mu_{\infty i}(1 + \alpha_i^2), \qquad \alpha_i = \frac{\gamma_{1i}}{\gamma_{2i}\rho}. \qquad (9.50)$$

It differs from the LQG index, as in Chapter 8, since it uses the H_∞ singular values instead of the LQG singular values, and it uses the additional multiplier α_i^2. The following properties of α_i are obtained. From Eq.(9.41b) one obtains $\kappa_i = \gamma_{2i}^2(1 - \alpha_i^2)$, therefore

$$\alpha_i \leq 1 \qquad \text{for} \quad \kappa_i \geq 0 \qquad (9.51a)$$

$$1 < \alpha_i^2 < 1 + \frac{1}{4\gamma_{1i}^2\gamma_{2i}^2} \qquad \text{for} \quad -\frac{1}{4\gamma_{1i}^2} < \kappa_i < 0 \qquad (9.51b)$$

The index $\sigma_{\infty i}$ serves as an indicator of importance of the ith balanced component of the H_∞ controller. If $\sigma_{\infty i}$ is small, the ith component is considered negligible and can be truncated.

It is easy to see that for an H_2 system, when $\rho^{-1} = 0$, one gets $\sigma_{\infty i} = \sigma_{2i}$, with

$$\sigma_{2i} = \mu_{2i}\gamma_{2i}^2 \qquad (9.52)$$

as introduced in Chapter 8.

The above choice of reduction index is justified by the properties of the closed-loop poles, presented below.

9.7.2 Closed-Loop Poles

Let $(A_\infty, B_\infty, C_\infty)$ be the state-space representation of the central H_∞ controller as in Eq.(9.15), i.e.

$$A_\infty = A + B_2 k_c + k_e C_2 + \rho^{-2} B_1 B_1^T S_{\infty c}, \qquad B_\infty = -k_e, \qquad C_\infty = k_c \qquad (9.53)$$

and $k_c=-B_2^T S_{\infty c}$, $k_e=-S_o S_{\infty e} C_2^T$, $S_o=(I-\rho^{-2}S_{\infty e}S_{\infty c})^{-1}$. Defining the closed-loop state variable $x_o^T=[x^T \ \varepsilon^T]$, where $\varepsilon=x-\hat{x}$, one obtains the closed-loop balanced state-space equations from Eqs.(9.12) and (9.53)

$$\dot{x}_o=A_o x_o+B_o w, \quad z=C_o x_o \tag{9.54a}$$

where

$$A_o=\begin{bmatrix} A_{11} & A_{12} \\ A_{21} & A_{22} \end{bmatrix}, \quad B_o=\begin{bmatrix} B_1 \\ B_1+k_e D_{21} \end{bmatrix}, \quad C_o=\begin{bmatrix} C_1+D_{12}k_c & -D_{12}k_c \end{bmatrix} \tag{9.54b}$$

$$A_{11}=A+B_2 k_c, \qquad A_{12}=-B_2 k_c, \tag{9.54c}$$

$$A_{21}=-\rho^{-2}B_1 B_1^T M_\infty, \qquad A_{22}=A+k_e C_2+\rho^{-2}B_1 B_1^T M_\infty \tag{9.54d}$$

Note that if

$$\sigma_{\infty i}\ll\sigma_{\infty 1}, \qquad \text{for } i=k+1,\dots,n, \tag{9.55}$$

then

$$A_{22i}\cong A_i-2\sigma_{\infty i}I_2, \tag{9.56}$$

i.e., the ith pole is shifted by $2\sigma_{\infty i}$ with respect to the open-loop location. To prove it, note that for flexible structures A is diagonally dominant, and the following components are diagonally dominant

$$B_2 k_c=B_2 B_2^T M_\infty\cong diag(2\zeta_i\omega_i \ \gamma_{2i}^2\mu_{\infty i})=diag(\frac{2\zeta_i\omega_i\sigma_{\infty i}}{\alpha_i^2}) \tag{9.57a}$$

$$k_e C_2=M_\infty C_2^T C_2\cong diag(2\zeta_i\omega_i \ \gamma_{2i}^2\mu_{\infty i})=diag(\frac{2\zeta_i\omega_i\sigma_{\infty i}}{\alpha_i^2}) \tag{9.57b}$$

220

$$\rho^{-2}B_1B_1^TM_\infty \cong diag(\frac{2\zeta_i\omega_i\gamma_{1i}^2\mu_{\infty i}}{\rho^2}) = diag(2\zeta_i\omega_i \; \sigma_{\infty i}), \qquad (9.57c)$$

thus each of four blocks of A_o is diagonally dominant. If $\sigma_{\infty i} \ll \sigma_{\infty 1}$ for $i = k+1,...,n$, then the ith diagonal components of A_{12} and A_{21} in the closed-loop matrix A_o (see Eq.(9.54)) are small for $i = k+1,...,n$. Thus for those components the separation principle is valid: gains k_{ci} and k_{ei} are independent. Furthermore, the ith diagonal block A_{22i} of the matrix A_{22} is as follows

$$A_{22i} \cong A_i - s_{oi}\mu_{\infty i}C_{2i}^TC_{2i} - \rho^{-2}B_{1i}B_{1i}^T\mu_{\infty i} \qquad (9.58)$$

where A_i is given by Eq.(9.32). Again, for $\sigma_{\infty i} \ll \sigma_{\infty 1}$ notice that $s_{oi} \cong 1$, that $\mu_{\infty i}C_{2i}^TC_{2i} \cong 2\zeta_i\omega_i\gamma_{1i}^2\mu_{\infty i}I_2$, and that $\rho^{-2}B_{1i}B_{1i}^T\mu_{\infty i} \cong 2\zeta_i\omega_i\rho^{-2}\gamma_{1i}^2\mu_{\infty i}I_2 = 2\zeta_i\omega_i\gamma_{1i}^2\mu_{\infty i}\alpha_i^2 I_2$. Consequently, Eq.(9.58) now becomes Eq.(9.56).

9.7.3 Controller Performance

Let the vector ε be partitioned as $\varepsilon^T = [\varepsilon_r^T, \; \varepsilon_t^T]$, with ε_r of dimension n_r, ε_t of dimension n_t, and $n_r + n_t = n$. Let the matrix of the reduction indices be arranged in decreasing order, $\Sigma_\infty = diag(\sigma_{\infty 1},..., \; \sigma_{\infty n})$, $\sigma_{\infty i} \geq \sigma_{\infty i+1}$, and be divided consistently with ε,

$$\Sigma_\infty = \begin{bmatrix} \Sigma_{\infty r} & 0 \\ 0 & \Sigma_{\infty t} \end{bmatrix}, \qquad (9.59)$$

where $\Sigma_{\infty r} = diag(\sigma_{\infty 1}, \; ... \; ,\sigma_{\infty k})$, $\Sigma_{\infty t} = diag(\sigma_{\infty k+1}, \; ... \; ,\sigma_{\infty n})$. Divide the matrix M_∞ accordingly, $M_\infty = diag(M_{\infty r}, M_{\infty t})$. The closed-loop system representation (A_o, B_o, C_o) is rearranged according to the division of ε.

$$A_o = \begin{bmatrix} A_{or} & A_{ort} \\ A_{otr} & A_{ot} \end{bmatrix}, \quad B_o = \begin{bmatrix} B_{or} \\ B_{ot} \end{bmatrix}, \quad C_o = \begin{bmatrix} C_{or} & C_{ot} \end{bmatrix} \tag{9.60}$$

Hence the closed-loop state is now $x_o^T = [x_r^T \ \varepsilon_t^T]$ and $x_r = [x \ \varepsilon_r]$. The reduced-order controller representation is (A_{or}, B_{or}, C_{or}), and let the closed-loop system state be denoted by $\bar{x}_r$.

For the condition of Eq.(9.55) satisfied, the performance of the closed-loop system with the reduced-order controller is almost identical to the full-order controller in the sense that $\|x_r - \bar{x}_r\| \cong 0$. It follows from Eq.(9.57) that for $\sigma_{\infty i} \ll \sigma_{\infty 1}$ $(i = k+1,\ldots,n)$ one obtains $\|A_{otr}\| \cong \|A_{ort}\| \cong 0$, and the closed-loop block A_{ot} is almost identical to the open-loop block A_t, i.e., $A_{ot} \cong A_t$. In this case, from Eqs.(9.54) and (9.60), one obtains

$$\dot{x}_r = A_{or}x_r + A_{ort}\varepsilon_t + B_{or}w \cong A_{or}x_r + B_{or}w = \dot{\bar{x}}_r \tag{9.61}$$

or, thus $x_r \cong \bar{x}_r$.

The above approximations hold for the controllers with the limited damping authority, i.e. for the controllers that modify only moderately the system natural frequencies. Typically, the states with largest H_∞ singular values do not fall under this category, but the states with the smallest H_∞ singular values are under the low authority control. Thus the latter states are exactly the most suitable for reduction, and therefore the presented reduction procedure is applicable anyway, and efficient, as it will be shown in the next section.

9.8 Design Examples

The H_∞ design method is illustrated with the simple flexible system, the truss structure, and for the Deep Space Network antenna.

9.8.1 Simple Structure

The design of an H_∞ controller for a system as in Fig.2.1 is described. The system parameters are as follows: $m_1=3$, $m_2=1$, $m_3=2$, $k_1=30$, $k_2=k_3=k_4=6$, $D=0.004K+0.001M$, where M, K, and D are mass, stiffness, and damping matrices, respectively. The control input (u) acts at mass No.2 and mass No.3 in opposite directions. The first disturbance (w_1) acts at mass No.2 and mass No.3 in opposite directions, with the amplification factor 3, the second disturbance (w_2) act at mass No. 2, and the third one (w_3) is the output noise. The output (y) is a displacement of mass No.2, and the controlled outputs (z_1, z_2, z_3) are the displacement of mass No.2 with the amplification factor 3, the rate of mass No.3, and the input u. Thus:

$$B_1=\begin{bmatrix} 0 & 0 & 0 \\ 0 & 0 & 0 \\ 0 & 0 & 0 \\ 0 & 0 & 0 \\ 1 & 3 & 0 \\ 0 & -1.5 & 0 \end{bmatrix}, \quad B_2=\begin{bmatrix} 0 \\ 0 \\ 0 \\ 0 \\ 1 \\ -0.5 \end{bmatrix}, \quad C_1=\begin{bmatrix} 0 & 0 & 1 & 0 & 0 & 0 \\ 0 & 0 & 0 & 0 & 3 & 0 \\ 0 & 0 & 0 & 0 & 0 & 0 \end{bmatrix}, \quad C_2=[0\ 0\ 0\ 0\ 1\ 0],$$

and $D_{12}^T=D_{21}=[0\ 0\ 1]$.

The parameter ρ is determined, obtaining $\rho=7.55$, and the H_∞ singular values are $M_\infty=diag(7.5473,\ 6.6567,\ 3.1142,\ 2.8568,\ 2.5929,\ 2.5877)$. The open- and closed-loop impulse responses are shown in Fig.9.7 (from the first input to the second output), and the magnitudes of transfer function of the open- and closed-loop systems are compared in Fig.9.8, showing a significant vibration suppression.

The controller is reduced, and the reduction index matrix is $\Sigma_\infty=diag(40.9502,\ 36.0697,\ 12.1002,\ 11.0862,\ 4.0827,\ 4.0702)$. The last two indices are small, hence the controller order is reduced from 6 to 4 states. The obtained closed-loop system is stable, with comparable performance, as is confirmed with the impulse responses of the full and the reduced controller in Fig.9.9 (from the first input to the second output).

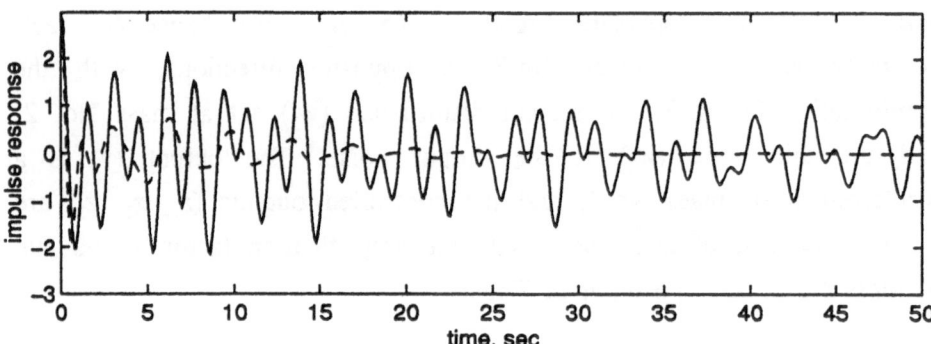

Figure 9.7: Impulse responses, open- and closed-loop systems, from the first input to the second output

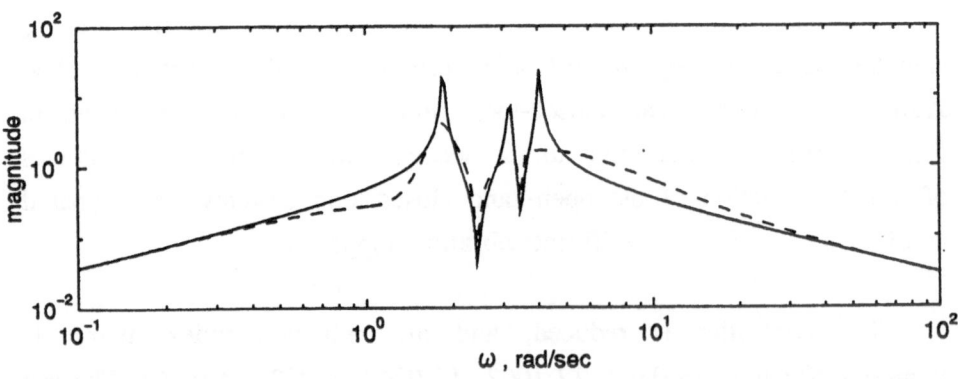

Figure 9.8: Magnitudes of transfer function, open- and closed-loop systems, from the first input to the second output

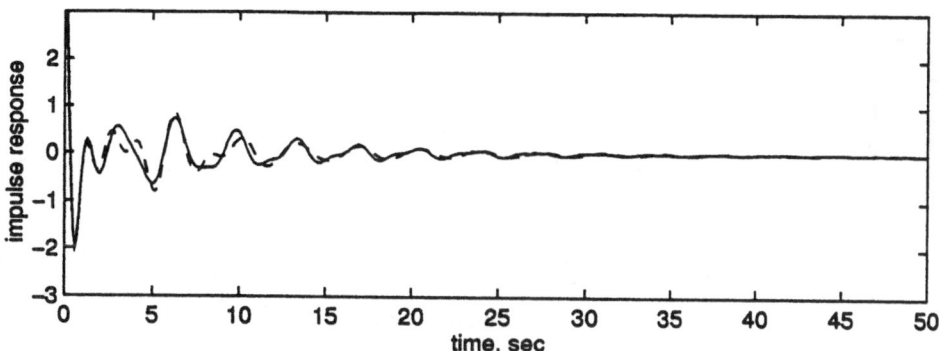

Figure 9.9: Impulse responses, full- and reduced-order controller, from the first input to the second output

9.8.2 Truss

The design of the H_∞ controller for the truss structure as shown in Fig.2.2 is presented. The structural model has $n=26$ states. All inputs and outputs are directed vertically. The disturbance w acts at node $n7$. The output z at node $n8$ is minimized. The controlled inputs u and outputs y are collocated at node $n4$, and the components of C_1 and B_1 are equal to 300 at node $n4$; other components are zero.

The system H_∞ characteristic values (solid line), H_2 characteristic values (dashed line), and Hankel singular values (dot-dashed line) are compared in Fig.9.10, showing that the relationship of Eq.(9.45) holds. The critical value is $\rho=469$.

In Fig.9.11 the H_∞ characteristic values obtained from Eq.(9.14) are compared with its approximate values computed from Eq.(9.42), showing that the approximate values are close to the exact ones. The H_∞ and H_2 reduction indices are shown in Fig.9.12; this figure shows that they coincide for $\sigma_{\infty i} \ll \sigma_{\infty 1}$.

225

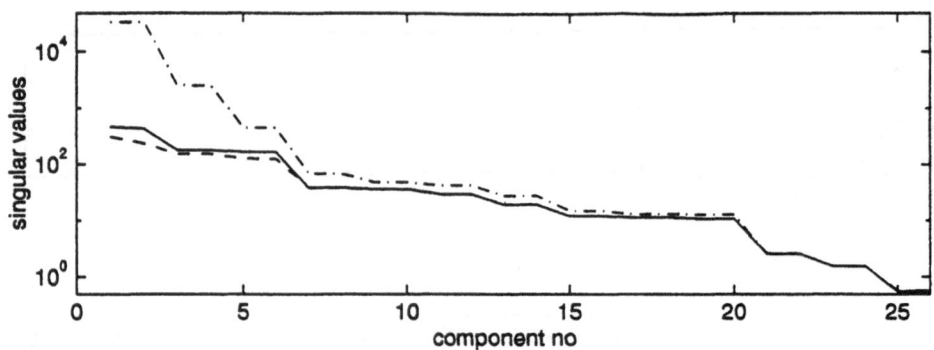

Figure 9.10: H_∞ (solid line), H_2 (dashed line), and Hankel singular values (dot-dashed line) of the truss

Open- and closed-loop impulse responses are compared in Fig.9.13. The H_∞ reduction index satisfies the condition in Eq.(9.55) for $k=8,\ldots,13$, i.e., $\sigma_{\infty k}<0.01 \ll \sigma_{\infty 1}$. Hence the controller can be reduced to 14 states. Indeed, the controller of order 14 is stable, and its performance is almost identical to the full-order controller (closed-loop impulse responses of the full- and reduced-order controllers overlap).

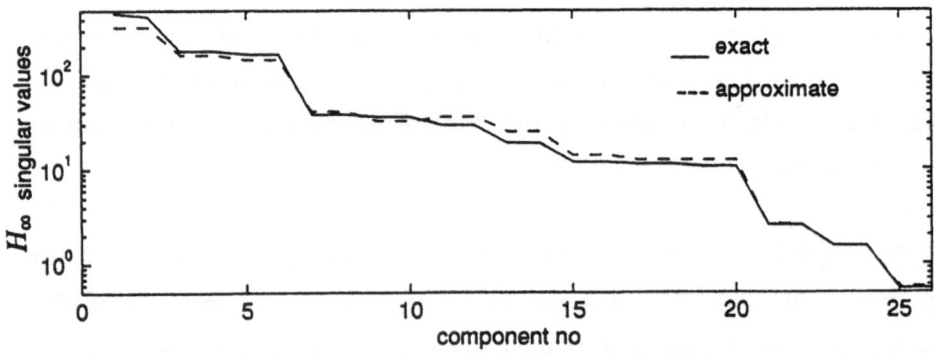

Figure 9.11: Exact and approximate H_∞ singular values

226

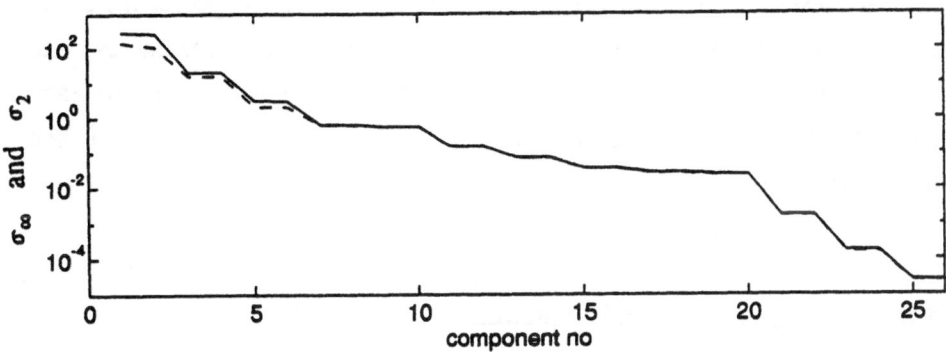

Figure 9.12: H_∞, H_2 reduction indices

The closeness of the open-loop and closed-loop balanced representations, as well as the modal representation, is estimated with the vectors ε_{h1i}, ε_{h2i}, and ε_{mi}, $i=1,...,13$, as defined in Eqs.(9.40) and (9.36). For the truss under consideration, the largest values of these vectors were typically in the range 0.98-0.99, with some in the range 0.9-0.98. The remaining values of the vectors were typically in the range 0-0.01, with some in the range 0.01-0.1, thus they satisfied

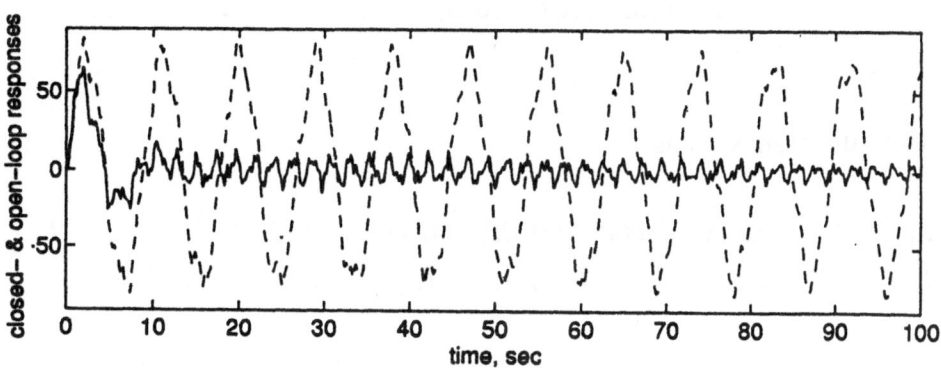

Figure 9.13: Open- and closed-loop system responses

227

the conditions in Eqs.(9.38) and (9.36). The closeness of the open- and closed-loop balanced representations, as well as modal representation, was also tested with simulations of the H_∞ controller performance in the H_∞ balanced coordinates, in the open-loop balanced coordinates, and in the modal coordinates. The results obtained were very close to each other for each set of coordinates, either for the full-order or the reduced-order controller.

9.8.3 Deep Space Network Antenna

The H_∞ controller for the Deep Space Network antenna is designed. The plant model has 35 states and is in the open-loop balanced representation. It is augmented with frequency filters at the exogenous input and regulated output as in Fig.9.14, to reflect wind disturbances and performance measures. However, instead of actually applying the filters, the plant balanced representation is modified as mentioned in Sec.9.4. The modifications include changes in the matrix B_1 (or alternatively C_1) to reflect the filter properties. The system matrix A remains unchanged. The modified matrix B_1 is as follows:

$B_1^T = 10^{-4} diag(63000, 63000, 100000, 100000, 2.8, 2.8, 1.6, 1.6, 0.9, 0.9,$
$0.5, 0.5, 0.3, 0.3, 0.2, 0.2, 0.12, 0.12, 0.07, 0.07, 0.04, 0.04, 0.03,$
$0.03, 0.015, 0.015, 0.01, 0.01, 0.005, 0.005, 0.003, 0.003, 0.002,$
$0.002, 0.001),$

while the matrix C_1 is

$C_1 = 10^{-4} diag(63000, 63000, 100000, 100000, 1, 1, 1, 1, 1, 1, 1, 1, 1, 1,$
$1, 1, 1, 1, 1, 1, 1, 1, 1, 1, 1, 1, 1, 1, 1, 1, 1, 1, 1, 1, 1).$

The first two terms of matrices B_1 and C_1 describe the participation of the integral of the position of the antenna (in azimuth and elevation) at the exogenous input and regulated output, respectively. The next two

228

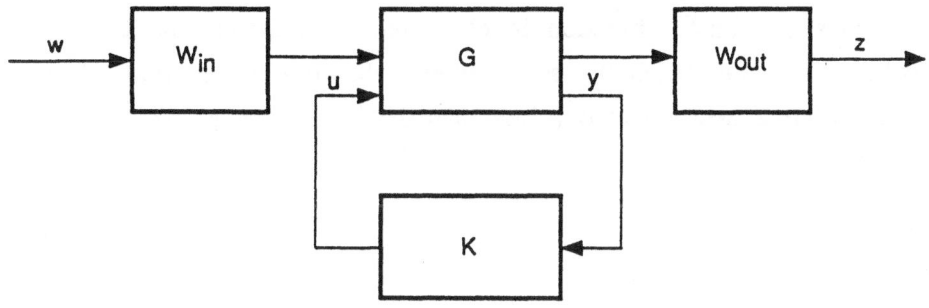

Figure 9.14: H_∞ control system with the frequency shaping filters

terms denote the participation of the position of the antenna (in azimuth and elevation). The remaining terms of B_1 are components of the frequency spectrum of the wind acting on the antenna. The remaining terms of C_1 are participants of higher frequency components in the exogenous output.

The first four terms in B_1 and C_1 (for position and integral of the position in azimuth and elevation) have been determined by shaping sequentially the step responses and the transfer functions: for good tracking, small acquisition time and small overshoot are required in the step response, and the magnitude of the transfer function should be equal to 1 for the given frequency bandwidth (in our case up to 2 Hz). The remaining components of C_1 are set equal to 0.0001, and were chosen to suppress excessive oscillatory motion of the antenna.

Matrix B_2 describes the elevation and azimuth rates, which are actuated inputs to the antenna, while C_2 describes encoder positions in elevation and azimuth, which are the outputs of the antenna.

The system was H_∞ balanced and controller gains were determined from Eq.(9.26), obtaining H_∞ parameter $\rho=27$. The step responses and magnitudes of transfer function in elevation and azimuth are shown in Figs.9.15a,b, and 9.16a,b, solid lines, showing good performance (settling time 2 sec, overshoot less than 10%, frequency bandwidth up to 2 Hz).

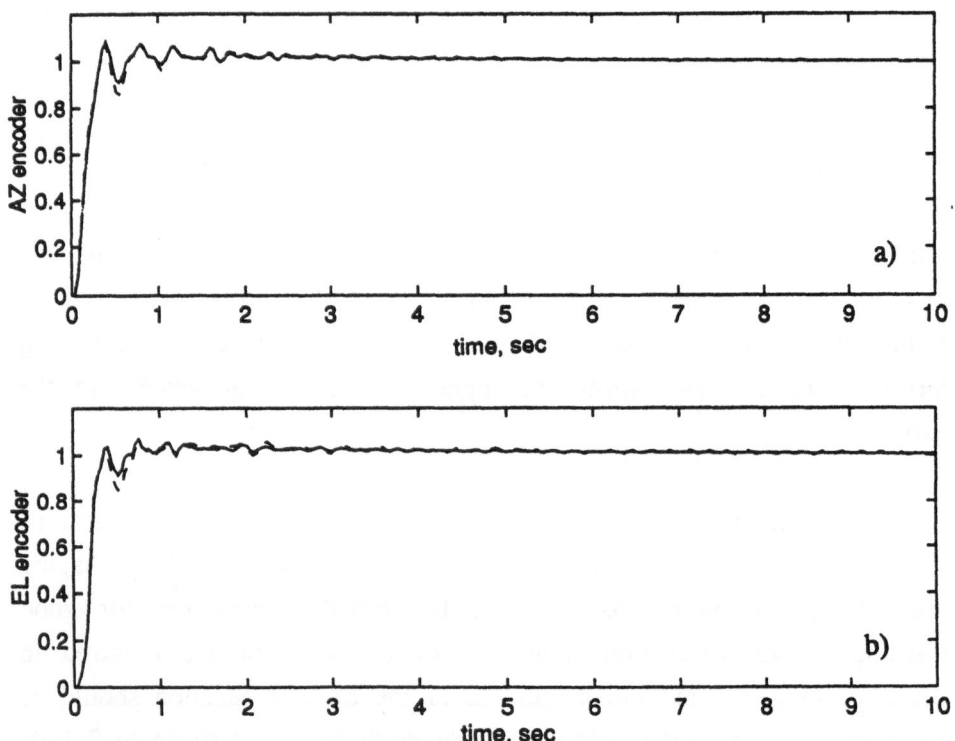

Figure 9.15: Step responses of closed-loop systems: (a) azimuth, (b) elevation

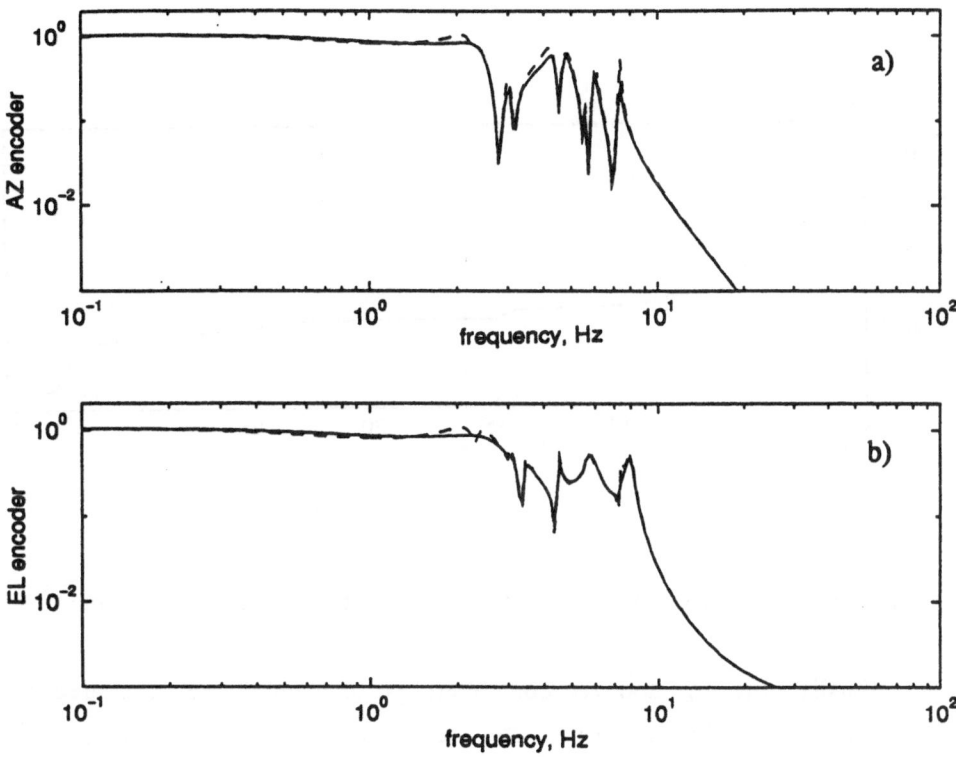

Figure 9.16: Magnitudes of transfer function of closed-loop systems: (a) azimuth, (b) elevation

The controller order of 35 was reduced, using the reduction indices $\sigma_{\infty i}$. The indices are plotted in Fig.9.17, showing large values for the four tracking components. The order 16 was chosen for the reduced controller. This reduced order controller stabilize the system, and the performance of this controller was comparable to the full-order controller, as shown in Figs.9.15, and 9.16, dashed lines.

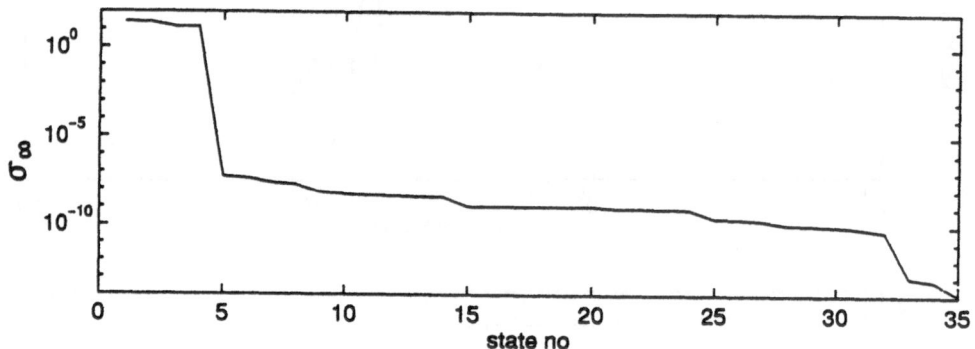

Figure 9.17: Reduction indices for the Deep Space Network antenna

Finally, the H_∞ and H_2 controllers performances are compared in Fig.9.18a,b, showing improved step response, and transfer function of the H_∞ controller when compared to the H_2 controller. Also, the wind disturbance suppression of the H_∞ controller improved by 25% when compared to the LQG controller from Chapter 8.

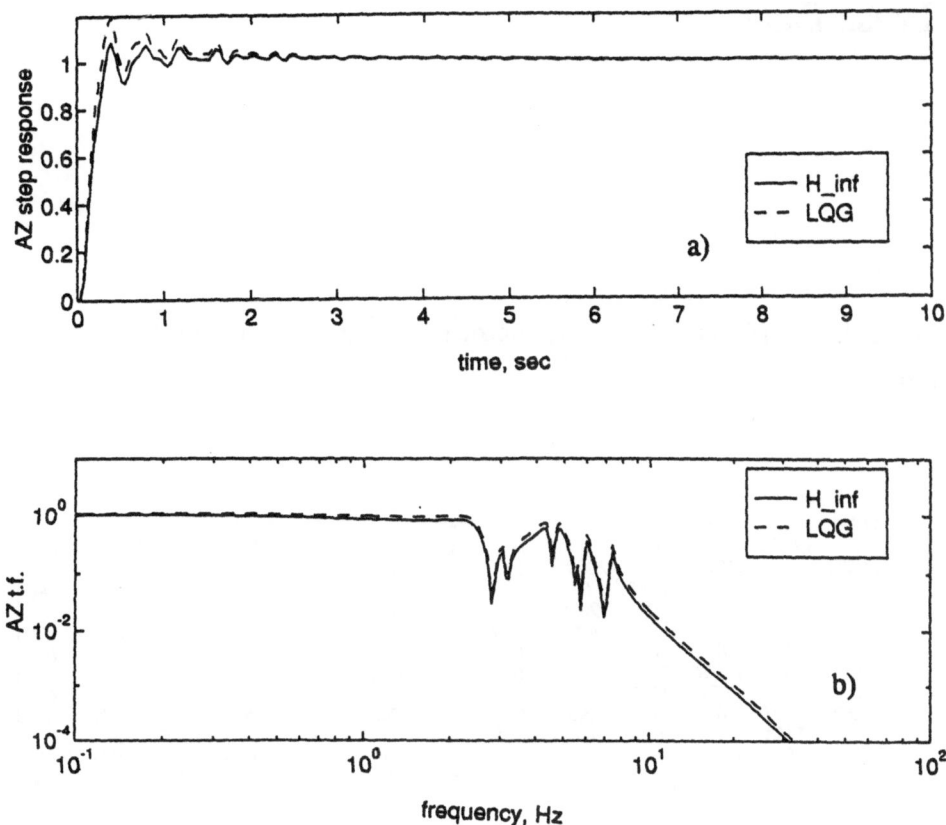

Figure 9.18: A comparison of the H_∞ and H_2 controllers for the Deep Space Network antenna: (a) step responses, (b) magnitudes of the transfer function

Appendix A

Truss Data

The stiffness (K) and mass (M) matrices of a truss as in Fig.2.2 are as follows:

$$K = \begin{bmatrix}
61993 & 0 & -20000 & 0 & 0 & 0 & 7697 & 0 & 0 & -10996 & -7697 & 0 & 0 \\
0 & 39348 & 0 & 0 & 0 & 0 & -5388 & 0 & -28571 & -7697 & -5388 & 0 & 0 \\
-20000 & 0 & 61993 & 0 & -20000 & 0 & 0 & -10996 & 7697 & 0 & 0 & -10996 & -7697 \\
0 & 0 & 0 & 39348 & 0 & 0 & 0 & 7697 & -5388 & 0 & -28571 & -7697 & -5388 \\
0 & 0 & -20000 & 0 & 30996 & -7697 & 0 & 0 & 0 & -10996 & 7697 & 0 & 0 \\
0 & 0 & 0 & 0 & -7697 & 33960 & 0 & 0 & 0 & 7697 & -5388 & 0 & -28571 \\
7697 & -5388 & 0 & 0 & 0 & 0 & 33960 & 0 & 0 & 0 & 0 & 0 & 0 \\
0 & 0 & -10996 & 7697 & 0 & 0 & 0 & 61993 & 0 & -20000 & 0 & 0 & 0 \\
0 & -28571 & 7697 & -5388 & 0 & 0 & 0 & 0 & 39348 & 0 & 0 & 0 & 0 \\
-10996 & -7697 & 0 & 0 & -10996 & 7697 & 0 & -20000 & 0 & 61993 & 0 & -20000 & 0 \\
-7697 & -5388 & 0 & -28571 & 7697 & -5388 & 0 & 0 & 0 & 0 & 39348 & 0 & 0 \\
0 & 0 & -10996 & -7697 & 0 & 0 & 0 & 0 & 0 & -20000 & 0 & 30996 & 7697 \\
0 & 0 & -7697 & -5388 & 0 & -28571 & 0 & 0 & 0 & 0 & 0 & 7697 & 33960
\end{bmatrix}$$

$$M = \begin{bmatrix} 514.1 & 514.1 & 514.1 & 514.1 & 292.1 & 292.1 & 292.1 & 514.1 & 514.1 & 514.1 & 514.1 & 292.1 & 292.1 \end{bmatrix}$$

Appendix B

Matlab Functions

Balanced open-loop systems as well as balanced LQG and H_∞ controllers are the backbones of the control system design methods presented in this book. For this reason the Matlab® functions, developed by the author, for the determination of the modal form 2 of the state-space representation *(realdiag)*, for the determination of the open-loop balanced representation *(bal_op_loop)*, for the determination of the open-loop balanced representation in the finite time and frequency intervals *(bal_time_freq)*, for the determination of the LQG balanced representation *(bal_LQG)*, and for the determination of the H_∞ balanced representation *(bal_H_inf)* are presented in this Appendix. Note that these functions use the following standard Matlab routines: *are, cdf2rdf, inv, lqe, lqr, lyap, norm, size, sqrt, svd.*

B1 Transformation to the Modal Form 2

% function [Am,Bm,Cm,Vm]=realdiag(A,B,C);
% This function finds the modal state-space representation of form 2
% The diagonal blocks of Am are in increasing order of imaginary part
% of the eigenvalues of A
% Vm is the transformation matrix from (A,B,C) to (Am,Bm,Cm)

```
function [Am,Bm,Cm,Vm]=realdiag(A,B,C);
%
[V,D]=eig(A);
[x,in]=sort(diag(abs(D)));
D=D(in,in);
V=V(in,in);
B=B(in,:);
C=C(:,in);
[Vm,Am]=cdf2rdf(V,D);
Bm=inv(Vm)*B;
Cm=C*Vm;
```

B2 Open-Loop Balanced Representation

```
% function [Ab,Bb,Cb,Gamma,R]=bal_op_loop(A,B,C);
% This function finds the open-loop balanced representation (Ab,Bb,Cb)
% so that controllability (Wc) and observability (Wo) grammians
% are equal and diagonal:
%                 Wc=Wo=Gamma
%
% Input parameters:
% (A,B,C)                - system state-space representation
%
% Output parameters:
% (Ab,Bb,Cb)             - balanced representation
% R                      - transformation to the balanced representation
% Gamma                  - Hankel singular values
%
%
```

```
function [Ab,Bb,Cb,Gamma,R]=bal_op_loop(A,B,C);
%
Wc=lyap(A,B*B');                    % controllability grammian
Wo=lyap(A',C'*C);                   % observability grammian
[Uc,Sc,Vc]=svd(Wc);                 % SVD of the controllability grammian
[Uo,So,Vo]=svd(Wo);                 % SVD of the observability grammian
Sc=sqrt(Sc);
So=sqrt(So);
P=Uc*Sc;
Q=So*Vo';
H=Q*P;                              % Hankel matrix
[V,Gamma2,U]=svd(H);                % SVD of the Hankel matrix
Gamma=sqrt(Gamma2);                 % Hankel singular values
R=P*U*inv(Gamma);                   % transformation matrix R
Rinv=inv(Gamma)*V'*Q;               % inverse of R
Ab=Rinv*A*R;
Bb=Rinv*B;
Cb=C*R;                             % balanced representation (Ab,Bb,Cb)
```

B3 Finite Time and Frequency Balanced Representation

```
% function [Ab,Bb,Cb,Gamma,R]=bal_time_freq(A,B,C,t1,t2,om1,om2 );
% This function finds the open-loop balanced representation (Ab,Bb,Cb)
% so that controllability (Wc) and observability (Wo) grammians
% are balanced in the time interval T=[t1,t2],
% and frequency interval OM=[om1,om2]
%
% Input parameters:
% (A,B,C)            - system state-space representation
% t1, t2             - lower and upper time limits
% om1, om2           - lower and upper frequency limits
%
```

237

```
% Output parameters:
% (Ab,Bb,Cb)              - balanced representation
% R                       - transformation to the balanced represent.
% Gamma                   - Hankel singular values
%
function [Ab,Bb,Cb,Gamma,R]=bal_time_freq(A,B,C,t1,t2,om1,om2);
%
% frequency transformation matrix Sw:
j=sqrt(-1);
[n1,n2]=size(A);
i=eye(n1);
x1=j*om1*i+A;
x2=inv(-j*om1*i+A);
Sw1=(j/2/pi)*logm(x1*x2);
x1=j*om2*i+A;
x2=inv(-j*om2*i+A);
Sw2=(j/2/pi)*logm(x1*x2);
Sw=Sw2-Sw1;
%
% infinite frequency grammians:
Wc=lyap(A,B*B');                   %controllability grammian
Wo=lyap(A',C'*C);                  %observability grammian
%
% finite frequency grammians:
Wcw=Wc*conj(Sw)'+Sw*Wc;
Wow=conj(Sw)'*Wo+Wo*Sw;
%
% finite time transformation matrices St1, St2:
St1=-expm(A*t1);
St2=-expm(A*t2);
```

```
% finite time and frequency grammians:
Wct1W=St1*Wcw*St1';
Wct2W=St2*Wcw*St2';
WcTW=Wct1W-Wct2W;
%
Wot1W=St1'*Wow*St1;
Wot2W=St2'*Wow*St2;
WoTW=Wot1W-Wot2W;
%
Wc=WcTW;
Wo=WoTW;
%
% balancing:
[Uc,Sc,Vc]=svd(Wc);          %SVD of contr. grammian
[Uo,So,Vo]=svd(Wo);          %SVD of observ. grammian
Sc=sqrt(Sc);
So=sqrt(So);
P=Uc*Sc;
Q=So*Vo';
H=Q*P;                       %Hankel matrix
[V,Gamma2,U]=svd(H);         %SVD of Hankel matrix
Gamma=sqrt(Gamma2);          %Hankel singular values
R=P*U*inv(Gamma);            %transformation matrix R
Rinv=inv(Gamma)*V'*Q;        %inverse of R
Ab=Rinv*A*R;
Bb=Rinv*B;
Cb=C*R;                      %balanced representation (Ab,Bb,Cb)
```

B4 LQG Balanced Representation

```
% function [Ab,Bb,Cb,Mu,Kpb,Keb,Qb,Vb,R]=bal_LQG(A,B,C,Q,R,V,W)
%
% This function finds the LQG balanced representation (Ab,Bb,Cb)
% so that CARE (Sc) and FARE (Se) solutions are equal and diagonal:
%                 Sc=Se=Mu
%
%
% Input parameters:
% (A,B,C)           - system state-space representation,
% Q                 - state weight matrix,
% R                 - input weight matrix,
% V                 - process noise covariance matrix,
% W                 - measurement noise covariance matrix.
%
% Output parameters:
% (Ab,Bb,Cb)        - LQG balanced representation,
% R                 - LQG balanced transformation,
% Mu                - balanced CARE, FARE solution
% Qb                - balanced weight matrix,
% Vb                - balanced process noise covariance matrix,
% Kpb, Keb          - balanced gains.
%
%
function [Ab,Bb,Cb,Mu,Kpb,Keb,Qb,Vb,R]=bal_LQG(A,B,C,Q,R,V,W)
%
V1=V;
R1=R;
[n1,n2]=size(A);
[Kp,Sc,ec]=lqr(A,B,Q,R);          % solution of CARE
[Ke,Se,ee]=lqe(A,eye(n1),C,V,W);  % solution of FARE
[Uc,Ssc,Vc]=svd(Sc);
```

```
Pc=sqrt(Ssc)*Vc';                          % Pc
[Ue,Sse,Ve]=svd(Se);
Pe=Ue*sqrt(Sse);                           % Pe
H=Pc*Pe;                                    % H
[V,Mu,U]=svd(H);                            % SVD of H
mu=sqrt(Mu);
R=Pe*U*inv(mu);                             % transformation R
Rinv=inv(mu)*V'*Pc;                         % inverse of R
Ab=Rinv*A*R;
Bb=Rinv*B;
Cb=C*R;                                     % LQG balanced representation
Qb=R'*Q*R;                                  % balanced weight matrix
Vb=Rinv*V1*Rinv';                           % balanced process noise cov. matrix
[Kpb,Scb,ecb]=lqr(Ab,Bb,Qb,R1);
[Keb,Seb,eeb]=lqe(Ab,eye(n1),Cb,Vb,W);     % balanced gains
```

B5 H_∞ Balanced Representation

```
% function [Ab,Bb1,Bb2,Cb1,Cb2,Mu_inf,R]=bal_H_inf(A,B1,B2,C1,C2,ro)
%
% This function finds the H_inf balanced representation
%          (Ab,Bb1,Bb2,Cb1,Cb2)
% so that HCARE (Sc) and HFARE (Se) solutions are equal and diagonal
%              Sc=Se=Mu_inf
%
% Input parameters:
% (A,B1,B2,C1,C2)            - system state-space representation
% ro                         - parameter in HCARE and HFARE
%
```

241

```
% Output parameters:
% (Ab,Bb1,Bb2,Cb1,Cb2)        - H_inf balanced representation,
% R                            - H_inf balanced transformation,
% Mu                           - balanced HCARE, HFARE solution,
%
%
function [Ab,Bb1,Bb2,Cb1,Cb2,Mu_inf,R]=bal_H_inf(A,B1,B2,C1,C2,ro)
%
[n1,n2]=size(A);
Qc=C1'*C1;
gi=1/(ro*ro);
Rc=B2*B2'-gi*B1*B1';
[Sc,sc1,sc2,wellposedc]=are(A,Qc,Rc,'eigen');   % HCARE   solution
Qe=B1*B1';
Re=C2'*C2-gi*C1'*C1;
[Se,se1,se2,wellposef]=are(A',Qe,Re,'eigen');   % HFARE solution
%
if(norm(imag(Se))>1e-6 | norm(imag(Sc))>1e-6) ...
disp('non-positive solution');end
%
[Uc,Ssc,Vc]=svd(Sc);
Pc=sqrt(Ssc)*Vc';                 % Pc
[Ue,Sse,Ve]=svd(Se);
Pe=Ue*sqrt(Sse);                  % Pe
N=Pc*Pe;
[V,Mu_inf,U]=svd(N);
mu_inf=sqrt(Mu_inf);
R=Pe*U*inv(mu_inf);
Rinv=inv(mu_inf)*V'*Pc;
Ab=Rinv*A*R;
Bb1=Rinv*B1;
Bb2=Rinv*B2;
Cb1=C1*R;
Cb2=C2*R;                                  % H_inf balanced representation
```

References

[1]   Aidarous, S.E., Gevers, M.R., and Installe, M.J., "Optimal
       Sensors' Allocation Strategies for a Class of Stochastic
       Distributed Systems," *Int. J. Control,* vol.22, 1975, pp.197-213.

[2]   Anderson, B.D.O., "A System Theory Criterion for Positive Real
       Matrices," *J. SIAM Control,* vol.5, 1967, pp.171-182.

[3]   Anderson, B.D.O., and Moore, J.B., *Optimal Control,* Prentice
       Hall, Englewood Cliffs, N.J., 1990.

[4]   Athans, M., "On the Design of PID Controllers Using Optimal
       Linear Regulator Theory," *Automatica,* vol.7, 1971, pp.643-647.

[5]   Aubrun, J.N., "Theory of the Control of Structures by Low-
       Authority Controllers," *Journal of Guidance, Control, and
       Dynamics,* vol.3, 1980.

[6]   Aubrun, J.N. and Margulies, G., "Low-Authority Control Synthesis
       for Large Space Structures," *NASA Contractor Report 3495,*
       Contract NAS1-14887, 1982.

[7]   Basseville, M., Benveniste, A., Moustakides, G.V., and Rougee,
       A., "Optimal Sensor Location for Detecting Changes in Dynamical
       Behavior," *IEEE Trans. Autom. Control,* vol.AC-32, 1987, pp.1067-
       1075.

[8]   Belvin, W. K., et al., "Langley's CSI Evolutionary Model: Phase
       0," NASA TM 104165, Langley Research Center, Hampton, VA., Nov.
       1991.

[9]   Benhabib, R.J., Iwens, R.P., and Jackson, R.L., "Stability of Large Space Structure Control Systems Using Positivity Concepts," *Journal of Guidance and Control*, vol.4, 1981, pp.487-494.

[10]   Blelloch, P.A., Mongori, D.L., and Wei, J.D., "Perturbation Analysis of Internal Balancing for Lightly Damped Mechanical Systems with Gyroscopic and Circulatory Forces," *Journal of Guidance, Control, and Dynamics*, vol.10, 1987, pp.406-410.

[11]   Boyd, S.P., and Barratt, C.H., *Linear Controller Design*, Prentice Hall, Englewood Cliffs, NJ, 1991.

[12]   Bruner, A., et al., "Active Vibration Absorber for the CSI Evolutionary Model: Design and Experimental Results," *Journal of Guidance, Control and Dynamics*, vol.15, 1992, pp.1253-1257.

[13]   Bryson, Jr., A.E., *Control of Spacecraft and Aircraft*, Princeton University Press, Princeton, NJ, 1994.

[14]   Ceton, C., Wortelboer, P.M.R., and Bosgra, O.H., "Frequency Weighted Closed-Loop Balanced Reduction," *European Control Conference*, 1993.

[15]   Clough, R. W., and Penzien, J., *Dynamics of Structures*, New York: McGraw Hill, 1975.

[16]   Colgren, R. D., "Methods for Model Reduction," *AIAA Guidance, Navigation and Control Conference*, Minneapolis, MN, 1988, pp.777-790.

[17]   Collins, Jr., E.G., Haddad, W.M., and Ying, S.S., "Construction of Low Authority, Nearly Non-Minimal LQG Compensators for Reduced-order Control Design," *1994 American Control Conference*, Baltimore, 1994.

[18] Cottin, N., Prells, U., and Natke, H.G. "On the Use of Modal Transformation to Reduce Dynamic Models in Linear System Identification," *IUTAM Symposium on Identification of Mechanical Systems*, Wuppertal, Germany, 1993.

[19] DeLorenzo, M.L., "Selection of Noisy Sensors and Actuators for Regulation of Linear Systems," Ph.D. Thesis, School of Aeronautics and Astronautics, Purdue University, West Lafayette, IN, May 1983.

[20] DeLorenzo, M.L., "Sensor and Actuator Selection for Large Space Structure Control," *Journal of Guidance, Control and Dynamics*, vol.13, 1990, pp.249-257.

[21] Derese, I., and Noldus, E., "Design of Linear Feedback Laws for Bilinear Systems," *Int. J. Control*, vol.31, 1980, pp.219-237.

[22] Desoer, C.A., and Vidyasagar, M., *Feedback Systems: Input-Output Properties*, Academic Press, New York, 1975.

[23] Dorf, R.C., *Modern Control Systems*, Addison-Wesley, Reading, MA, 1980.

[24] Doyle, J. C., Glover, K., Khargonekar, P.P., and Francis, B.A., "State Space Solutions to Standard H_2 and H_∞ Control Problems," *IEEE Trans. Automatic Control*, vol.34, 1989, pp.831-847.

[25] Doyle, J.C., B.A. Francis, and A.R. Tannenbaum, *Feedback Control Theory*, Macmillan, NY, 1992.

[26] Enns, D.F., "Model Reduction for Control System Design," Ph.D. Thesis, Dept. Aeronautics and Astronautics, Stanford University, Palo Alto, CA.

[27] Ewins, D.J., *Modal Testing*. Wiley, New York, 1984.

[28] Francis, B.A., and Wonham, W.M., "The Internal Model Principle of Control Theory," *Automatica*, vol.12, 1976, pp.457-465.

[29] Fukata, S., Mohri, A., and Takata, M., "On the Determination of the Optimal Feedback Gains for Multivariable Linear Systems Incorporating Integral Action," *Int. J. Control*, vol.31, 1980, pp.1027-1040.

[30] Furuta, K., A. Sano, and D. Atherton, *State Variable Methods in Automatic Control*. Wiley, Chichester, 1988.

[31] Gawronski, W., "Model Reduction for Flexible Structures: Test Data Approach," *Journal of Guidance, Control, and Dynamics*, vol.14, 1991, pp.692-694.

[32] Gawronski, W., "Balanced Systems and Structures: Reduction, Assignment, and Perturbations," in: *Control and Dynamics Systems*, ed. C.T. Leondes, vol.54, Academic Press, San Diego, 1992, pp.373-415.

[33] Gawronski, W., "Linear Quadratic Controller Design for the Deep Space Network Antennas," *Journal of Guidance, Control, and Dynamics*, vol.17, 1994, pp.655-660.

[34] Gawronski, W., "Balanced LQG Compensator for Flexible Structures," *Automatica*, vol.30, 1994, pp.1555-1564.

[35] Gawronski, W., and Hadaegh, F.Y., "Balanced Input-Output Assignment," *International Journal for Systems Science*, vol.24, 1993, pp.1027-1036.

[36] Gawronski, W., and Juang, J.N., "Near-Optimal Model Reduction in Balanced or Modal Coordinates," *26th Annual Allerton Conference on Communication, Control, and Computing*, Monticello, 1988, pp.209-218.

[37] Gawronski, W., and Juang, J.N., "Model Reduction in Limited Time and Frequency Intervals," *Int. Journal of Systems Science*, vol.21, 1990, pp.349-376.

246

[38] Gawronski, W., and Juang, J.N., "Model Reduction for Flexible Structures," in: *Control and Dynamics Systems*, ed. C.T. Leondes, vol.36, Academic Press, San Diego, 1990, pp.143-222.

[39] Gawronski, W., and Lim, K.B., "Controllability and Observability of Flexible Structures with Proof-Mass Actuators," *Journal of Guidance, Control, and Dynamics*, vol.16, 1993, pp.899-902.

[40] Gawronski, W., and Lim, K.B., "Balanced H_∞ and H_2 Controllers," *Proc. 1994 IEEE American Control Conference*, Baltimore, MD, 1994, pp.1116-1122.

[41] Gawronski, W., and Lim, K.B., "Balanced Actuator and Sensor placement for Flexible Structures," *Proc. AIAA Guidance, Navigation, and Control Conference*, Baltimore, MD, 1995.

[42] Gawronski, W., and Mellstrom, J.A., "Model Reduction for Systems with Integrators," *Journal of Guidance, Control, and Dynamics*, vol.15, 1992, pp.1304-1306.

[43] Gawronski, W., and Mellstrom, J.A., "Control and Dynamics of the Deep Space Network Antennas," a chapter in: *Control and Dynamics Systems*, ed. C.T. Leondes, vol.63, Academic Press, San Diego, 1994, pp.289-412.

[44] Gawronski, W., and Natke, H.G., "Balanced State Space Representation in the Identification of Dynamic Systems," a chapter in *Application of System Identification in Engineering*, ed. H.G. Natke, Springer, New York, 1988.

[45] Gawronski W., and Natke, H.G., "Realizations of the Transfer Function Matrix," *Int. Journal of Systems Science*, vol.18, 1987, pp.229-236.

[46] Gawronski, W., Racho, C., and Mellstrom, J.A., "Application of the LQG and Feedforward Controllers to the Deep Space Network Antennas," *IEEE Trans. on Control System Technology*, 1995.

[47] Gawronski, W., and Williams, T., "Model Reduction for Flexible Space Structures," *Journal of Guidance, Control, and Dynamics,* vol.14, No.1, 1991, pp.68-76.

[48] Glover, K., "All Optimal Hankel-Norm Approximations of Linear Multivariable Systems and their L^{∞}-Error Bounds," *International Journal of Control*, vol.39, 1984, pp.1115-1193.

[49] Glover, K., and Doyle, J.C., "State-Space Formulae for All Stabilizing Controllers That Satisfy an H_{∞}-norm Bound and Relations to Risk Sensitivity," *Systems and Control Letters,* vol.11, 1988, pp.167-172.

[50] Golub, G.H., Van Loan, C.F., *Matrix Computations*, The Johns Hopkins University Press, Baltimore, MD, 1989.

[51] Gregory, Jr., C.Z., "Reduction of Large Flexible Spacecraft Models Using Internal Balancing Theory," *J. Guidance, Control and Dynamics*, vol.7, 1984, pp.725-732.

[52] Hamdan, A.M.A., and Nayfeh, A.H., "Measures of Modal Controllability and Observability for First- and Second-Order Linear Systems," *Journal of Guidance, Control and Dynamics*, vol.12, 1989, pp.421-428, and p.768.

[53] Ho, B.L., and Kalman, R.E., "Effective Construction of Linear State-Variable Models From Input-Output Data," *Regelungstechnik*, vol.14, 1966, pp.545-548.

[54] Horn, R.A., and Johnson, C.R., *Matrix Analysis*, Cambridge University Press, Cambridge, England, 1985.

[55] Hyland, D.C., Bernstein, D.S., "The Optimal Projection Equations for Model Reduction and the Relationships among the Methods of Wilson, Skelton, and Moore," *IEEE Trans. on Autom. Control*, vol.AC-30, 1985, pp.1201-1211.

248

[56] Hyland, D.C., "Distributed Parameter Modeling of Flexible Spacecraft: Where's the Beef?" *NASA Workshop on Distributed Parameter Modeling and Control of Flexible Aerospace Systems,* Williamsburg, VA, 1992, NASA Conference Publication 3242, 1994, pp.3-23.

[57] Inman D.J., *Vibration with Control, Measurements, and Stability,* Prentice Hall, Englewood Cliffs, NJ, 1989.

[58] Johnson, C.D., "Optimal Control of the Linear Regulator with Constant Disturbances," *IEEE Trans. Automatic Control,* vol.13, 1968, pp.416-421.

[59] Jonckheere, E.A., and Silverman, L.M., "Singular Value Analysis of Deformable Systems," *20th IEEE Conference on Decision and Control,* San Diego, 1981, pp.660-668.

[60] Jonckheere, E.A., and Silverman, L.M., A New Set of Invariants for Linear Systems - Application to Reduced Order Controller Design. *IEEE Trans. Autom. Control.* vol.AC-28, 1983, pp.953-964.

[61] Jonckheere, E.A., "Principal Component Analysis of Flexible Systems - Open Loop Case," *IEEE Trans. Autom. Control,* vol.27, 1984, pp.1095-1097.

[62] Joshi, S.M., *Control of Large Flexible Space Structures,* Springer, Berlin, 1989.

[63] Juang, J-N., and Rodriguez, G., "Formulations and Applications of Large Structure Actuator and Sensor Placements," *VPI/AIAA Symp. on Dynamics and Control of Large Flexible Spacecraft,* Blacksburg, 1979, pp.247-262.

[64] Juang, J.-N., *Applied System Identification,* Prentice Hall, Englewood Hills, NJ, 1994.

[65] Junkins, J.L., editor, *Mechanics and Control of Large Flexible Structures*. Progress in Astronautics and Aeronautics, vol.129, AIAA, Washington, DC, 1990.

[66] Junkins, J.L., and Kim, Y., *Introduction to Dynamics and Control of Flexible Structures*, Published by the American Institute of Aeronautics and Astronautics, Washington, D.C., 1993.

[67] Kailath, T., *Linear Systems*, Prentice Hall, Englewood Cliffs, NJ, 1980.

[68] Kammer, D., "Sensor Placement for On-Orbit Modal Identification and Correlation of Large Space Structures," *Journal of Guidance, Control and Dynamics*, vol.14, 1991, pp.251-259.

[69] Kim, Y., and Junkins, J.L., "Measure of Controllability for Actuator Placement," *Journal of Guidance, Control and Dynamics*, vol.14, 1991, pp.895-902

[70] Kwakernaak, H., and Sivan, R., *Linear Optimal Control Systems*, Wiley, New York, 1972.

[71] Levine, W.S., and Reichert, R.T., "An Introduction to H_∞ Control Design," *Proc. 29th Conference on Decision and Control*, Honolulu, Hawaii, 1990, pp.2966-2974.

[72] Li, X.P., and Chang, B.C., "On Convexity of H_∞ Riccati Solutions and its Applications," *IEEE Trans. Autom. Control*, vol.38, 1993, pp.963-966.

[73] Lim, T.W., "Sensor Placement for On-Orbit Modal Testing," *32nd Structures, Structural Dynamics, and Materials Conf.* Baltimore, 1991.

[74] Lim, K.B., "Method for Optimal Actuator and Sensor Placement for Large Flexible Structures," *Journal of Guidance, Control and Dynamics*, vol.15, 1992, pp.49-57.

[75] Lim, K.B., Maghami, P.G., and Joshi, S.M., "Comparison of Controller Designs for an Experimental Flexible Structure," *IEEE Control Systems Magazine*, June 1992, pp.108-118.

[76] Lim, K.B., and Balas, G.J., "Line-of-Sight Control of the CSI Evolutionary Model: μ Control," *IEEE American Control Conference*, Chicago, IL, 1992.

[77] Lim, K.B., and Gawronski, W., "Actuator and Sensor Placement for Control of Flexible Structures," a chapter in: *Control and Dynamics Systems*, ed. C.T. Leondes, vol.57, Academic Press, San Diego, 1993, pp.109-152.

[78] Lin, C.F., *Advanced Control Systems Design*, Prentice Hall, Englewood Cliffs, 1994.

[79] Lindberg Jr., R.E., and Longman, R.W., "On the Number and Placement of Actuators for Independent Modal Space Control," *J. Guidance, Control and Dynamics*, vol.7, 1984, pp.215-223.

[80] Ljung, L., *System Identification*, Prentice Hall, Englewood Cliffs, 1987.

[81] Longman, R.W., and Horta, L.G., "Actuator Placement by Degree of Controllability Including the Effect of Actuator Mass," *Proceedings of the Seventh VPI&SU/AIAA Symposium on Dynamics and Control of Large Flexible Spacecraft*, Blacksburg, VA, 1989.

[82] Longman, R. W., and Alfriend, K. T., "Energy Optimal Degree of Controllability and Observability for Regulator and Maneuver Problems," *The Journal of the Astronautical Sciences*, vol.38, 1990, pp.87-103.

[83] Luenberger, D. G., *Optimization by Vector Space Methods*, New York: Wiley, 1969

[84] Maciejowski, J.M., *Multivariable Feedback Design*, Addison-Wesley, Wokingham, 1989.

[85] Maghami, P.G., and Joshi, S.M., "Sensor/Actuator Placement for Flexible Space Structures," *IEEE American Control Conference*, San Diego, CA, 1990, pp.1941-1948.

[86] Maghami, P.G., and Joshi, S.M., "Sensor-Actuator Placement for Flexible Structures with Actuator Dynamics," *AIAA Guidance, Navigation, and Control Conference*, New Orleans, LA, 1991, pp.46-54.

[87] McLaren, M.D., and Slater, G.L., "Robust Multivariable Control of Large Space Structures Using Positivity," *Journal of Guidance, Control, and Dynamics*, vol.10, 1987, pp.393-400.

[88] Meirovitch, L. *Dynamics and Control of Structures*, Wiley, New York, 1990.

[89] Moore, B.C., "Principal Component Analysis in Linear Systems, Controllability, Observability and Model Reduction," *IEEE Trans. Autom. Control*, vol.26, 1981, pp.17-32.

[90] Mustafa, D., H_∞-Characteristic Values, *28th IEEE Conference on Decision and Control*, Tampa, FL, 1988, pp.1483-1487.

[91] Mustafa, D., and Glover, K., "Controller Reduction by H_∞-Balanced Truncation," *IEEE Trans. Autom. Control*, vol.36, 1991, pp.668-682.

[92] Natke, H.G., *Einfuehrung in die Theorie und Praxis der Zeitreihen- und Modalanalyse*. Vieweg, Braunschweig, 1992.

[93] Opdenacker, P., and Jonckheere, E.A., LQG Balancing and Reduced LQG Compensation of Symmetric Passive Systems," *Int. J. Control*, vol.41, 1985, pp.73-109.

252

[94] Parnebo, L., and Silverman, L. M., "Model Reduction via Balanced State Space Representation," *IEEE Transactions on Automatic Control*, vol.AC-27, 1982, pp.382-387.

[95] Popov, V.M., *Hyperstability of Automatic Control Systems*, Springer, New York, 1973.

[96] Porter, B., "Optimal Control of Multivariable Linear Systems Incorporating Integral Feedback," *Electronics Letters*, vol.7, 1971.

[97] Porter, B., and Crossley R., *Modal Control*, Taylor and Francis, London, 1972.

[98] Racho, C., and Gawronski, W., "Experimental Modification and Identification of the DSS-13 Antenna Control System," *TDA Progress Report*, 42-115, Jet Propulsion Laboratory, California Institute of Technology, pp.42-53.

[99] Salama, M., Rose, T., and Garba, J., "Optimal Placement of Excitations and Sensors for Verification of Large Dynamical Systems," *28th Structures, Structural Dynamics and Materials Conference*, Monterey, CA, 1987, pp.1024-1031.

[100] Shahian, B., and Hassul, M., *Control System Design Using Matlab*, Prentice Hall, Englewood Cliffs, 1993.

[101] Skelton, R.E., and Hughes, P.C., "Modal Cost Analysis for Linear Matrix Second-Order Systems," *Journal of Dynamic Systems, Measurements and Control*, vol.102, 1980, pp.151-158.

[102] Skelton, R.E., "Cost Decomposition of Linear Systems with Application to Model Reduction," *Int. Journal of Control*, vol.32, 1980, pp.1031-1055.

[103] Skelton, R.E., and Yousuff, A., "Component Cost Analysis of Large Scale Systems," *Int. Journal of Control*, vol.37, 1983, pp.285-304.

[104] Skelton, R.E., and DeLorenzo, M.L., "Selection of Noisy Actuators and Sensors in Linear Stochastic Systems," *Journal of Large Scale Systems, Theory and Applications*, vol.4, 1983, pp.109-136.

[105] Skelton, R.E., *Dynamic System Control: Linear System Analysis and Synthesis*, Wiley, New York, 1988.

[106] Skelton, R.E., Singh, R., and Ramakrishnan, J., "Component Model Reduction by Component Cost Analysis," *AIAA Guidance, Navigation and Control Conference*, Minneapolis, 1988, pp.264-2747.

[107] Slater, G.L., Bosse, A.B., and Zhang, Q., "Robustness of Positive Real Controllers for Large Space Structures," *Journal of Guidance, Control, and Dynamics*, vol.15, 1992, pp.58-64.

[108] Stoorvogel, A., *The H_∞ Control Problem*, Prentice Hall, New York, 1992.

[109] Viswanathan, C. N., Longman, R. W., and Likins P. W., "A Degree of Controllability Definition: Fundamental Concepts and Application to Modal Systems," *Journal of Guidance, Control, and Dynamics*, vol.7, 1984, pp.222-230.

[110] Voth, C.T., Richards Jr., K.E., Schmitz, E., Gehling, R.N., and Morgenthaler, D.R., "Integrated Active and Passive Control Design Methodology for the LaRC CSI Evolutionary Model," *NASA Contractor Report 4580*, Contract NAS1-19371, 1994.

[111] Wen, J.T., "Time Domain and Frequency Domain Conditions for Strict Positive Realness," *IEEE Trans. on Automat. Control*, vol.AC-33, 1988 p.988-992.

[112] Wicks, M.A., and DeCarlo, R.A., "Grammian Assignment of the Lyapunov Equation," *IEEE Trans. on Automat. Control,* vol.AC-35, 1990, pp.465-.

[113] Willems, J.C., "Dissipative Dynamical Systems, Part I: General Theory, Part II: Linear Systems with Quadratic Supply Rates," *Arch. Rational Mech. and Anal.,* vol.45, 1972, pp.321-351, and pp.352-392.

[114] Willems, J.C., "Realization of Systems with Internal Passivity and Symmetry Constraints," *Journal of the Franklin Institute,* vol.301, 1976, pp.605-620.

[115] Williams, T., "Closed Form Grammians and Model Reduction for Flexible Space Structures," *IEEE Transactions on Automatic Control,* vol.AC-35, 1990, pp.379-382.

[116] Wilson, D.A., "Optimum Solution of Model-Reduction Problem," *IEE Proceedings,* vol.119, 1970, pp.1161-1165.

[117] Wilson, D.A., "Model Reduction for Multivariable Systems," *International Journal of Control,* vol.20, 1974, pp.57-60.

[118] Wortelboer, P., and van Oostveen, H., "Modal Reduction Guided by Hankel Singular Value Intervals," *Selected Topics in Identification, Modeling and Control,* vol.1, 1990.

[119] Wortelboer, P., and Bosgra, O.H., "Generalized Frequency Weighted Balanced Reduction," *Selected Topics in Identification, Modeling and Control,* vol.5, 1992

[120] Wortelboer, P., "Frequency-Weighted Balanced Reduction of Closed-Loop Mechanical Servo Systems: Theory and Tools," Ph. D. Dissertation, Technische Universiteit Delft, Holland, 1994.

[121] Yahagi, T., "Optimal Output Feedback Control in the Presence of Step Disturbances," *Int. J. Control,* vol.26, 1977, pp.753-762.

[122] Young, P.C., and Willems, J.C., "An Approach to the Linear Multivariable Servomechanism Problem," *Int. J. Control*, vol.15, 1972, pp.961-979.

[123] Zimmerman, D.C, and Inman, D.J., "On the Nature of the Interaction Between Structures and Proof-Mass Actuators," *J. Guidance, Control, and Dynamics*, vol.13, 1990, pp.82-88.

Index

D

E

F

G

H

Lecture Notes in Control and Information Sciences

Edited by M. Thoma

1992–1995 Published Titles:

Vol. 180: Kall, P. (Ed.)
System Modelling and Optimization.
Proceedings of the 15th IFIP Conference,
Zurich, Switzerland, September 2-6, 1991
969 pp. 1992 [3-540-55577-3]

Vol. 181: Drane, C.R.
Positioning Systems - A Unified Approach
168 pp. 1992 [3-540-55850-0]

Vol. 182: Hagenauer, J. (Ed.)
Advanced Methods for Satellite and Deep
Space Communications. Proceedings of
an International Seminar Organized by
Deutsche Forschungsanstalt für Luft-und
Raumfahrt (DLR), Bonn, Germany,
September 1992
196 pp. 1992 [3-540-55851-9]

Vol. 183: Hosoe, S. (Ed.)
Robust Control. Proceesings of a Workshop
held in Tokyo, Japan, June 23-24, 1991
225 pp. 1992 [3-540-55961-2]

Vol. 184: Duncan, T.E.; Pasik-Duncan, B.
(Eds)
Stochastic Theory and Adaptive Control.
Proceedings of a Workshop held in
Lawrence, Kansas, September 26-28,
1991
500 pp. 1992 [3-540-55962-0]

Vol. 185: Curtain, R.F. (Ed.); Bensoussan,
A.; Lions, J.L.(Honorary Eds)
Analysis and Optimization of Systems:
State and Frequency Domain Approaches
for Infinite-Dimensional Systems.
Proceedings of the 10th International
Conference, Sophia-Antipolis, France, June
9-12, 1992.
648 pp. 1993 [3-540-56155-2]

Vol. 186: Sreenath, N.
Systems Representation of Global Climate
Change Models. Foundation for a Systems
Science Approach.
288 pp. 1993 [3-540-19824-5]

Vol. 187: Morecki, A.; Bianchi, G.;
Jaworeck, K. (Eds)
RoManSy 9: Proceedings of the Ninth
CISM-IFToMM Symposium on Theory and
Practice of Robots and Manipulators.
476 pp. 1993 [3-540-19834-2]

Vol. 188: Naidu, D. Subbaram
Aeroassisted Orbital Transfer: Guidance
and Control Strategies
192 pp. 1993 [3-540-19819-9]

Vol. 189: Ilchmann, A.
Non-Identifier-Based High-Gain Adaptive
Control
220 pp. 1993 [3-540-19845-8]

Vol. 190: Chatila, R.; Hirzinger, G. (Eds)
Experimental Robotics II: The 2nd
International Symposium, Toulouse,
France, June 25-27 1991
580 pp. 1993 [3-540-19851-2]

Vol. 191: Blondel, V.
Simultaneous Stabilization of Linear
Systems
212 pp. 1993 [3-540-19862-8]

Vol. 192: Smith, R.S.; Dahleh, M. (Eds)
The Modeling of Uncertainty in Control
Systems
412 pp. 1993 [3-540-19870-9]

Vol. 193: Zinober, A.S.I. (Ed.)
Variable Structure and Lyapunov Control
428 pp. 1993 [3-540-19869-5]

Vol. 194: Cao, Xi-Ren
Realization Probabilities: The Dynamics of
Queuing Systems
336 pp. 1993 [3-540-19872-5]

Vol. 195: Liu, D.; Michel, A.N.
Dynamical Systems with Saturation
Nonlinearities: Analysis and Design
212 pp. 1994 [3-540-19888-1]

Vol. 196: Battilotti, S.
Noninteracting Control with Stability for
Nonlinear Systems
196 pp. 1994 [3-540-19891-1]

Vol. 197: Henry, J.; Yvon, J.P. (Eds)
System Modelling and Optimization
975 pp approx. 1994 [3-540-19893-8]

Vol. 198: Winter, H.; Nüßer, H.-G. (Eds)
Advanced Technologies for Air Traffic Flow
Management
225 pp approx. 1994 [3-540-19895-4]

Vol. 199: Cohen, G.; Quadrat, J.-P. (Eds)
11th International Conference on
Analysis and Optimization of Systems –
Discrete Event Systems: Sophia-Antipolis,
June 15–16–17, 1994
648 pp. 1994 [3-540-19896-2]

Vol. 200: Yoshikawa, T.; Miyazaki, F. (Eds)
Experimental Robotics III: The 3rd
International Symposium, Kyoto, Japan,
October 28-30, 1993
624 pp. 1994 [3-540-19905-5]

Vol. 201: Kogan, J.
Robust Stability and Convexity
192 pp. 1994 [3-540-19919-5]

Vol. 202: Francis, B.A.; Tannenbaum, A.R.
(Eds)
Feedback Control, Nonlinear Systems,
and Complexity
288 pp. 1995 [3-540-19943-8]

Vol. 203: Popkov, Y.S.
Macrosystems Theory and its Applications:
Equilibrium Models
344 pp. 1995 [3-540-19955-1]

Vol. 204: Takahashi, S.; Takahara, Y.
Logical Approach to Systems Theory
192 pp. 1995 [3-540-19956-X]

Vol. 205: Kotta, U.
Inversion Method in the Discrete-time
Nonlinear Control Systems Synthesis
Problems
168 pp. 1995 [3-540-19966-7]

Vol. 206: Aganovic, Z.;.Gajic, Z.
Linear Optimal Control of Bilinear Systems
with Applications to Singular Perturbations
and Weak Coupling
133 pp. 1995 [3-540-19976-4]

Vol. 207: Gabasov, R.; Kirillova, F.M.;
Prischepova, S.V. (Eds)
Optimal Feedback Control
224 pp. 1995 [3-540-19991-8]

Vol. 208: Khalil, H.K.; Chow, J.H.;
Ioannou, P.A. (Eds)
Proceedings of Workshop on Advances in
Control and its Applications
300 pp. 1995 [3-540-19993-4]

Vol. 209: Foias, C.; Özbay, H.;
Tannenbaum, A.
Robust Control of Infinite Dimensional
Systems: Frequency Domain Methods
230 pp. 1995 [3-540-19994-2]

Vol. 210: De Wilde, P.
Neural Network Models: An Analysis
164 pp. 1996 [3-540-19995-0]